HYDROGEN ENERGY
and VEHICLE SYSTEMS

GREEN CHEMISTRY AND CHEMICAL ENGINEERING

Series Editor: Sunggyu Lee
Ohio University, Athens, Ohio, USA

Proton Exchange Membrane Fuel Cells: Contamination and Mitigation Strategies
Hui Li, Shanna Knights, Zheng Shi, John W. Van Zee, and Jiujun Zhang

Proton Exchange Membrane Fuel Cells: Materials Properties and Performance
David P. Wilkinson, Jiujun Zhang, Rob Hui, Jeffrey Fergus, and Xianguo Li

Solid Oxide Fuel Cells: Materials Properties and Performance
Jeffrey Fergus, Rob Hui, Xianguo Li, David P. Wilkinson, and Jiujun Zhang

Efficiency and Sustainability in the Energy and Chemical Industries: Scientific Principles and Case Studies, Second Edition
Krishnan Sankaranarayanan, Jakob de Swaan Arons, and Hedzer van der Kooi

Nuclear Hydrogen Production Handbook
Xing L. Yan and Ryutaro Hino

Magneto Luminous Chemical Vapor Deposition
Hirotsugu Yasuda

Carbon-Neutral Fuels and Energy Carriers
Nazim Z. Muradov and T. Nejat Veziroğlu

Oxide Semiconductors for Solar Energy Conversion: Titanium Dioxide
Janusz Nowotny

Lithium-Ion Batteries: Advanced Materials and Technologies
Xianxia Yuan, Hansan Liu, and Jiujun Zhang

Process Integration for Resource Conservation
Dominic C. Y. Foo

Chemicals from Biomass: Integrating Bioprocesses into Chemical Production Complexes for Sustainable Development
Debalina Sengupta and Ralph W. Pike

Hydrogen Safety
Fotis Rigas and Paul Amyotte

Biofuels and Bioenergy: Processes and Technologies
Sunggyu Lee and Y. T. Shah

Integrated Biorefineries: Design, Analysis, and Optimization
Paul R. Stuart and Mahmoud M. El-Halwagi

Hydrogen Energy and Vehicle Systems
Scott E. Grasman

GREEN CHEMISTRY AND CHEMICAL ENGINEERING

HYDROGEN ENERGY
and VEHICLE SYSTEMS

Edited

Scott E. Grasman

CRC Press
Taylor & Francis Group
Boca Raton London New York

CRC Press is an imprint of the
Taylor & Francis Group, an **informa** business

CRC Press
Taylor & Francis Group
6000 Broken Sound Parkway NW, Suite 300
Boca Raton, FL 33487-2742

First issued in paperback 2017

© 2013 by Taylor & Francis Group, LLC
CRC Press is an imprint of Taylor & Francis Group, an Informa business

No claim to original U.S. Government works

Version Date: 2012920

ISBN 13: 978-1-4398-2681-2 (hbk)
ISBN 13: 978-1-138-07173-5 (pbk)

Library of Congress Cataloging-in-Publication Data

Hydrogen energy and vehicle systems / [edited by] Scott E. Grasman.
   pages cm. -- (Green chemistry and chemical engineering)
   Includes bibliographical references and index.
   ISBN 978-1-4398-2681-2 (hardback)
   1. Hydrogen as fuel. 2. Fuel cells. 3. Hybrid electric cars. I. Grasman, Scott E. (Scott Erwin), editor of compilation.

TP359.H8H935 2013
665.8'1--dc23
                                                             2012025560

Visit the Taylor & Francis Web site at
http://www.taylorandfrancis.com

and the CRC Press Web site at
http://www.crcpress.com

Contents

Section III   Hydrogen Safety

Series Preface

The subjects and disciplines of chemistry and chemical engineering have encountered a new landmark in the way of thinking about, developing, and designing chemical products and processes. This revolutionary philosophy, termed *green chemistry and chemical engineering*, focuses on the designs of products and processes that are conducive to reducing or eliminating the use and/or generation of hazardous substances. In dealing with hazardous or potentially hazardous substances, there may be some overlap with and inter-relationship between environmental chemistry and green chemistry. While environmental chemistry is the chemistry of the natural environment and pollutant chemicals in nature, green chemistry proactively aims to reduce and prevent pollution at its very source. In essence, the philosophies of green chemistry and chemical engineering tend to focus more on industrial application and practice rather than academic principles and phenomeno-logical science. However, as both a chemistry and chemical engineering phi-losophy, green chemistry and chemical engineering derives from and builds upon organic chemistry, inorganic chemistry, polymer chemistry, fuel chem-istry, biochemistry, analytical chemistry, physical chemistry, environmental chemistry, thermodynamics, chemical reaction engineering, transport phe-nomena, chemical process design, separation technology, automatic process control, and more. In short, green chemistry and chemical engineering is the rigorous use of chemistry and chemical engineering for pollution prevention and environmental protection.

The Pollution Prevention Act of 1990 in the United States established a national policy to prevent or reduce pollution at its source whenever fea-sible. In adhering to the spirit of this policy, the Environmental Protection Agency (EPA) launched its Green Chemistry Program in order to promote innovative chemical technologies that reduce or eliminate the use or genera-tion of hazardous substances in the design, manufacture, and use of chemi-cal products. Global efforts in green chemistry and chemical engineering have recently gained substantial support from the international commu-nities of science, engineering, academia, industry, and governments in all phases and aspects.

Some of the successful examples and key technological developments include the use of supercritical carbon dioxide as a green solvent in separa-tion technologies; application of supercritical water oxidation for destruction of harmful substances; process integration with carbon dioxide sequestration steps; solvent-free synthesis of chemicals and polymeric materials; exploita-tion of biologically degradable materials; use of aqueous hydrogen peroxide for efficient oxidation; development of hydrogen proton exchange membrane (PEM) fuel cells for a variety of power generation needs; advanced biofuel productions; devulcanization of spent tire rubber; avoidance of the use of chemicals and processes causing generation of volatile organic compounds (VOCs); replacement of traditional petrochemical processes by microorganism-based bioengineering processes; replacement of chlorofluorocarbons (CFCs) with nonhazardous alternatives; advances in design of energy efficient processes; use of clean, alternative, and renewable energy sources in manufacturing; and much more. This list, even though it is only a partial compilation, is undoubtedly growing exponentially.

This book series on green chemistry and chemical engineering by CRC Press/Taylor & Francis Group is designed to meet the new challenges of the 21st century in the chemistry and chemical engineering disciplines by publishing books and monographs based upon cutting-edge research and development to the effect of reducing adverse impacts upon the environment by chemical enterprise. In achieving this, the series will detail the development of alternative sustainable technologies that will minimize the hazard and maximize the efficiency of any chemical choice. The series aims at delivering the readers in academia and industry with an authoritative information source in the field of green chemistry and chemical engineering. The publisher and series editor are fully aware of the rapidly evolving nature of the subject and its long-lasting impact upon the quality of human life in both the present and future. As such, the team is committed to making this series the most comprehensive and accurate literary source in the field of green chemistry and chemical engineering.

Sunggyu Lee

Foreword

The gasoline crisis way back in the 1970s produced an initial awareness on the part of the general public about the finiteness of popular energy sources. Since that time, enthusiasm by the public, investors, and government administrations has surged and waned for various technologies viewed as possible "silver bullet" answers to the energy needs of developed and developing nations. In recent years, a more realistic notion has been adopted by many that there probably are no technology silver bullets but rather that an "all of the above" technological approach is needed.

Hydrogen technology can be one part of a comprehensive energy approach. While hydrogen enthusiasts tout the inherent cleanness of this basic element found prolifically in nature, detractors like to point out the challenges and costs of producing, storing, and using hydrogen on a large scale. While the debate between enthusiasts and detractors influences the level of interest by governments and the public, thankfully there are engineers and scientists in academia and industry who continue researching, developing, and applying hydrogen technologies toward ever more practical solutions.

This book provides insights from many of those scientists and engineers on a broad array of issues, challenges, and accomplishments of hydrogen technology over the past years leading to the present. As for the future for hydrogen technology, whether the glass is half full or half empty may be a matter of perspective. However, hydrogen technology seems certain to have a role in our energy future.

William R. Taylor

Preface

Purpose and Audience

Hydrogen shows great promise both as an energy carrier and as a fuel for transportation, portable, and stationary sources; however, the expanded use of hydrogen as a renewable energy source raises a number of concerns and challenges that complicate planning efforts. Organizations are researching, developing, and validating hydrogen pathways to establish a business case for market implementation. However, significant research and educational challenges still must be addressed.

The use of hydrogen technologies addresses critical societal issues related to energy security, stability, and sustainability. First, hydrogen may be produced from local resources, thus eliminating the need for complex energy/fuel supply lines. Second, hydrogen, used in conjunction with renewable energy sources, provides a stable method of energy/fuel production. Third, hydrogen produced from renewable sources provides clean, emission-free energy/fuel.

Hydrogen technology constitutes a highly interdisciplinary field that extends from the fundamentals of materials, electrochemical processes, and fuel processing/storage systems, to complex design concepts for hybrid vehicles, and renewable power/fuel systems. Infrastructure analysis, market transformation, public policy, safety, and environment also play key roles. Additionally, sustainable energy systems is an emerging field that aims to develop new and improved energy technologies, systems, and services, while understanding the impact of energy on the economy and society.

Hydrogen technology research encompasses traditional engineering disciplines (biological, chemical, electrical, environmental, geological, material science, mechanical, systems), sciences (biology, chemistry, mathematics, physics), social sciences (economics, psychology), and business. Thus, the book addresses transformational interdisciplinary research in the emerging field of sustainable energy systems to disprove common misconceptions regarding hydrogen technologies and demonstrate that hydrogen technologies are a viable part of a sustainable, stable, and secure energy infrastructure.

The book addresses a new comprehensive approach to the applications of hydrogen-based technologies aimed at integrating the transportation and electric power generation sectors to improve the efficiency and reliability of both systems. Improving the overall efficiency and performance of any/all stages will decrease costs and improve market penetration, which is critical to the long-term success of hydrogen technologies. The book also addresses intelligent energy management schemes for hydrogen energy and vehicle systems, as well as safety and environmental science related to hydrogen technologies and the infrastructure required to provide for safe, renewable-hydrogen options.

Major themes of this book are focused on hydrogen fuel and fuel cell technologies (including safety and environmental science), hydrogen vehicle systems, hydrogen energy systems, and hydrogen infrastructure and marketing strategies. Whereas other books focus on specific aspects of hydrogen (e.g., materials, fuel processing, fuel cell electrochemistry), this book aims to be a comprehensive look at state-of-the-art research in hydrogen energy and vehicle systems.

There is strong interest in hydrogen as an energy carrier that has gained support from industrial companies and a continuously growing level of government backing. While the establishment of a sustainable hydrogen economy is seen as key to long-term environmental and economic stability, achieving these societal benefits involves a variety of stakeholders on regional, national, and global levels. Thus, this book will do the following:

- Provide the basis for pursuing a broad research agenda to develop, demonstrate, evaluate, and promote the long-term successful use of hydrogen-based technologies
- Develop resources to attain, coordinate, and articulate stable and independent energy benefits

This book is appropriate for researchers and professionals in energy-related fields, faculty in related disciplines, students in related majors, and policy makers. It may be used as a reference for the practitioner or for university courses, short courses, and workshops. For example, courses are being taught as part of integrated energy curriculum in over 250 related programs. These courses have titles such as Energy Systems, Alternative Energy/Fuels, Hydrogen Systems, Fuel Cell Applications, Automotive Fuel Cell Systems, and Renewable Systems.

Authored by experts in the field, the book clearly and accurately presents a comprehensive resource on hydrogen systems. It provides a balanced presentation of hydrogen technology from both theoretical/technical and application perspectives. Based on current research in hydrogen energy and vehicle systems, it connects hydrogen technology through proper systems analysis and integration, including both quantitative and qualitative factors, and includes all stakeholder perspectives, including energy and environmental perspectives of hydrogen technologies.

Overview of the Book

The chapters published in this book are authored by more than 25 researchers affiliated with higher education institutions and/or research centers.

The chapters cover a range of theory and application and are grouped into three sections.

Section I: Hydrogen Energy and Fuel Cell Modeling

Chapter 1: Hydrogen and Electricity: Parallels, Interactions, and Convergence

This chapter discusses some of the major ways that a future hydrogen economy would interact with the electricity sector and how the transportation and stationary fuels sectors and the electricity sectors might converge. H_2 and electricity are both zero-carbon, flexible, useful, and complementary energy carriers that could provide power for a wide range of applications. Hydrogen is touted as an important future transportation fuel in the light-duty sector because of its storage characteristics, efficiency, and emissions. In addition, an important consideration for the evolution of the energy system is the competition and synergies for the use of energy resources for producing H_2 and electricity.

Chapter 2: Hydrogen Infrastructure: Production, Storage, and Transportation

This chapter provides a review of hydrogen production, storage, and transportation technologies. The production technologies selected represent promising near- and long-term options based upon state of technology, scale of production quantities, and environmental impacts. The production technologies include steam methane reformation, gasification, electrolysis, and thermochemical conversion. Compressed gas, liquid, cryo-compressed, metal hydride, and surface adsorption storage methods are presented based on the scale of hydrogen storage capability. The chapter concludes with a discussion on transportation methods and operational characteristics for an expanded hydrogen infrastructure.

Chapter 3: PEM Fuel Cell Basics and Computational Modeling

This chapter discusses the operational principles of polymer electrolyte membrane (PEM) fuel cells and presents models incorporated into a commercial software package. The models are evaluated against independent data reported in the literature for its suitability to predict the performance of PEM fuel cells. The findings establish a model capable of simulating PEM fuel cells with a reasonable degree of accuracy and the low computational intensity inherent to analytical modeling. Given the software environment the model is implemented in, this could be of significant aid to the design and optimization of fuel cell– and hybrid-powered vehicles.

Chapter 4: Dynamic Modeling and Control of PEM Fuel Cell Systems

This chapter discusses the basic principles of fuel cells including the history and different types of fuel cells along with their properties, structure, and applications, with a special focus on polymer electrolyte membrane (PEM) fuel cells.

Auxiliary devices needed for safe and efficient operation of PEM fuel cells are also introduced. Some well-known control-oriented dynamic models of PEM fuel cell components are presented. Simulation analysis of a typical PEM fuel cell is conducted based on the dynamic control-oriented models. Finally, commonly used control algorithms, such as oxygen excess ratio and temperature regulation, are presented and implemented using the control-oriented models.

Section II: Market Transformation and Applications

Chapter 5: Market Transformation Lessons for Hydrogen from the Early History of the Manufactured Gas Industry

This chapter explores the future for hydrogen by delving into the history of the manufactured gas industry, drawing comparisons and contrasts, and highlighting potentially valuable analogies and lessons. It examines various side-by-side comparisons between the two energy systems, including physical and chemical properties, costs, production processes, and system configurations, and examines infrastructure developments over time, reviewing five major phases in the history of manufactured gas. It concludes with five key analogies or lessons for hydrogen based upon this historical review.

Chapter 6: Fuel Cell Technology Demonstration and Data Analysis

This chapter strives to provide an independent third-party technology assessment that focuses on fuel cell system and hydrogen infrastructure performance, operation, maintenance, and safety. U.S. government-funded hydrogen and fuel cell demonstrations support technology research and development, and researchers at the National Renewable Energy Laboratory (NREL) are working to validate hydrogen and fuel cell systems in real-world settings. A key component of these demonstrations and deployments involves data collection, analysis, and reporting. NREL's Hydrogen Secure Data Center (HSDC) was established in 2004 as the central location for data analysis and works with the DOE and its fuel cell award teams to collect and analyze data from these early deployment and demonstration projects. The analysis is regularly updated and published by application and is summarized in this chapter.

Chapter 7: Producing Hydrogen for Vehicles via Fuel Cell–Based Combined Heat, Hydrogen, and Power: Factors Affecting Energy Use, Greenhouse Gas Emissions, and Cost

This chapter introduces the concept of producing fuel for hydrogen-powered vehicles using combined heat, hydrogen, and power (CHHP) systems based on stationary high-temperature fuel cells, which also provide electricity and heat to buildings. In addition, it explores the factors affecting the performance of CHHP systems in various locations as well as the associated greenhouse gas (GHG) emissions and hydrogen cost. The energy, GHG, and hydrogen cost implications of this technological strategy for facilitating efforts to establish a fueling infrastructure to support early hydrogen vehicle markets have been modeled; the analysis employs the FCPower model, which was developed by the National Renewable Energy Laboratory and is available for download as an Excel spreadsheet. This chapter explains some of the basic modeling assumptions underlying the representation of molten carbonate fuel cell (MCFC) systems in the FCPower model and reviews the total energy use, emissions, and hydrogen cost for CHHP installations in comparison to conventional supplies of energy to buildings and small-scale dedicated (SMR) production of hydrogen.

Chapter 8: Hybrid and Plug-In Hybrid Electric Vehicles

The introduction of high-power and high-energy dense lithium-ion-based electrochemical storage technologies has provided the necessary transformative advance to bring forth the recent focus on hybrid and plug-in hybrid electric vehicles. Hybridization of hydrogen combustion and hydrogen fuel cell propulsion systems can provide benefits similar to those seen with conventional vehicles. This chapter discusses the benefits and consequences of the various hybrid electric vehicle propulsion architectures as they are applied to hydrogen propulsion technology. These architectures have varying benefits to the propulsion system based on their ability to influence the output power of the vehicle relative to the goal of the vehicle hybridization. Further, the difference between hybrid and plug-in hybrid electric vehicles represents a varying degree of energy storage that must be considered with the size, class, and intent of the vehicle propulsion system. The appropriate application of the hybrid vehicle architecture with an accompanying energy management control will be crucial to the advancement of these vehicles.

Chapter 9: Hydrogen as Energy Storage to Increase Wind Energy Penetration into Power Grid

This chapter presents an analysis of a full wind and hydrogen integration. This study is a part of the activities carried out in the framework of the IEA Hydrogen Agreement, Task 24 "Wind Energy and Hydrogen Integration." As a result of the study, it is concluded that hydrogen could compete with other energy storage systems, mainly in energy applications linked to renewable energies such as wind power. Hydrogen can be stored for a future reconversion into electricity or can be used in a different application, taking advantage of its energy vector feature. Although there are still several disadvantages and technical problems to be solved, it presents optimism concerning the future of hydrogen in the energy sector. The chapter also discusses applications for hydrogen such as CHP and CHHP.

Chapter 10: Hydrogen Design Case Studies

This chapter discusses real-world applications of hydrogen technologies for a hydrogen community. The applications are generic and are applicable for communities around the world. They include a commercial hydrogen fueling station, residential hydrogen fueling, hydrogen applications for airports, and other hydrogen applications. These conceptual designs were created by the Missouri University of Science and Technology's hydrogen student design team in response to the Fuel Cell Hydrogen Energy Association (formerly known as National Hydrogen Association) Hydrogen Student Design Contests.

Section III: Hydrogen Safety

Chapter 11: Hydrogen Safety

This chapter discusses hydrogen safety in a "hydrogen infrastructure" setting such as hydrogen fueling stations, hydrogen vehicle research and development garages, hydrogen storage, and stationary fuel cell installations—each with different risks and potential hazards. Hydrogen has many properties that make it unique including wide flammability limits, low ignition energy, high diffusivity, and low flame visibility. With proper understanding of these properties, incorporating experience, and safe handling procedures, hydrogen can be used in a safe working environment.

Chapter 12: Hydrogen Fuel Cell Vehicle Regulations, Codes, and Standards

This chapter covers regulations, codes, and standards (RCS) for hydrogen fuel cell vehicles. The chapter covers both domestic vehicle standards found primarily in Society of Automotive Engineers (SAE) and CSA Standards (CSA) documents and international standards found primarily in International Organization for Standardization (ISO) standards. The chapter does not cover the motor vehicle safety regulations promulgated by federal transportation safety agencies outside of the United States. The basic purpose of these RCS is to ensure safe operation of fuel cell–powered vehicles. These RCS do not cover the infrastructure required to support these vehicles. The infrastructure requirements are well developed in the United States, but they are outside of the scope of this chapter.

MATLAB® is a registered trademark of The Math Works, Inc. For product information, please contact: The Math Works, Inc., 3 Apple Hull Drive, Natick, MA. Tel. +508-647-7000; Fax: +508-647-7001; E-mail: info@mathworks.com; Web: http://www.mathworks.com

Acknowledgments

In addition to the authors of the chapters, the completion of this book has involved the efforts of several people.

The editor is grateful to anonymous reviewers for their insightful comments and suggestions for improvements to individual chapters. The chapter submissions were subject to a blind refereeing process that engaged up to three reviewers per chapter. Without these efforts, this book could not have been completed. The editor is particularly grateful to series editor Sunggyu "KB" Lee, Fermin Mallor, and William Taylor for assistance in the development of the book.

Last, but not least, the editor would like to acknowledge the help of the CRC/Taylor & Francis team, particularly Amber Donley and Allison Shatkin, as their assistance and patience have been significant.

Scott E. Grasman
Industrial and Systems Engineering, Rochester Institute of Technology

About the Editor

Scott E. Grasman is a professor and head of Industrial and Systems Engineering Department at Rochester Institute of Technology. He has had previous appointments in engineering management and systems engineering at Missouri University of Science and Technology, operations and manufacturing management in the Olin Business School at Washington University in St. Louis, statistics and operations research at the Public University of Navarre and Universitat Oberta de Catalunya. He received BSE, MSE, and PhD degrees in industrial and operations engineering from the University of Michigan. He has relevant industrial experience, including collaborations on research and curriculum activities.

His primary research interests relate to the application of quantitative models, focusing on the design and development of supply chain and logistics networks. Dr. Grasman has been a principal or co-principal investigator on projects with support from, among others, the Air Force Research Lab, Argonne National Labs, Army Engineer Research and Development Center, Bi-State Development Agency, Boeing, Defense Logistics Agency, Ford, General Motors, Government of Spain, Intel Research Council, Missouri DoT, Missouri Research Board, NSF, SAP America, Semiconductor Research Corporation, TranSystems, U.S. Department of Education, U.S. Department of Energy, U.S. Department of State, U.S. Department of Transportation, Walmart Logistics, and others. His work has resulted in being the author or coauthor of over 100 technical papers, including multiple best conference paper awards, as well as reviewer/editorial roles for various technical journals and books.

Dr. Grasman has significant expertise in the areas of operations research, management science, and supply chain and logistics systems. Within these areas, he has developed mathematical models that aid managerial decision making by generating theoretical and applied solutions to important problems. He has provided solutions for random yield production systems, manufacturing processes (e.g., headcount allocation, cross-training, and scheduling), enterprise integration (e.g., integrated inventory/transportation systems and collaborative SMEs), and information sharing through connective technologies. Recent and on-going studies have focused on alternative fuels programs, public–private partnerships in transportation, alternative energy infrastructure modeling/simulation, and sustainability in supply chain and facility logistics. His research also addresses energy/engineering education. Dr. Grasman is a member of ASEE, IIE, and INFORMS. His email is Scott.Grasman@rit.edu.

About the Contributors

Mónica Aguado is a manager of the Renewable Energy Grid Integration Department within the National Renewable Energy Centre (CENER) in Sarriguren, Spain. She holds a PhD in industrial engineering from Universidad Publica de Navarra and has 15 years of experience as a researcher and engineer in both the private and public sector. She is an expert on power electrical systems, mainly on two aspects: grid integration of renewable energies and aspects related to electromagnetic transients in power systems. She is also a professor of electric and electronic engineering at the Public University of Navarra in Pamplona, Spain. Dr. Aguado is the author of numerous scientific publications and has participated in a large number of national and international expert groups and committees.

Leslie Eudy is a senior project leader at NREL. Her work involves evaluating alternative fuel, hybrid, and fuel cell propulsion technologies in heavy-duty applications. She currently works in NREL's Hydrogen Technologies and Systems Center as part of the hydrogen technology validation team. Her projects focus on collecting and analyzing operational and performance data on hydrogen and fuel cell buses to help determine the status of the technology in real-world service.

Gabriel García is an industrial engineering researcher at the Renewable Energy Grid Integration Department (IRE) within the National Renewable Energy Centre (CENER) in Sarriguren, Spain. He obtained the industrial engineer degree at Basque Country University in 2002. He developed his career in the private sector and joined CENER in 2005. He is an expert on the integration of renewables and energy storage systems (ESSs) and has participated in several private projects related to hydrogen and other ESSs. García is the author of several scientific publications and conferences and has also participated in national and international projects, expert groups, and committees.

Raquel Garde has a PhD in inorganic chemistry from Universidad de Zaragoza and 18 years of international experience in the field of chemistry and physics (catalysis, magnetism, solid state, etc.). She has developed most of her outstanding research career at the university (Germany and France) and at the National Renewable Energy Centre (CENER) in Sarriguren, Spain. Since 2002, she has been responsible for the Energy Storage Group within the Renewable Energy Grid Integration Department (IRE). She works mainly on hydrogen fuel cells and electrochemical energy storage systems, as well as energy management with electric vehicles, cold warehouses, and so on.

Dr. Garde is the author of numerous scientific publications and conferences and has participated in a large number of national and international projects, expert groups, and committees.

Joseph Ishaku is currently a student pursuing a master's degree in the Department of Mechanical and Aerospace Engineering at the Missouri University of Science and Technology. He received his BS degree in mechanical engineering from the Missouri University of Science and Technology in 2009. He worked as the Controls Team leader on the EcoCAR project organized by Argonne National Labs. His research interests include modeling, simulation, and control of mechatronics systems, hydrogen PEM fuel cells, lithium-ion batteries, and integration and control of hybrid electric systems.

Umit O. Koylu is a professor of mechanical and aerospace engineering at the Missouri University of Science and Technology (Missouri S&T). He holds PhD and MS degrees from the University of Michigan at Ann Arbor and a BS degree from Istanbul Technical University, all in aerospace engineering. His research includes conventional energy (combustion, IC engines, coal power plants, fluid and heat transport), alternative energy (characterization of solid oxide fuel cells, modeling of proton exchange membrane fuel cells, hydrogen technologies, alternative fuels, clean coal technologies), environmental science and technology (air pollutants, particulates, measurement techniques), and various fields (synthesis of nanoparticles, laser diagnostics, fire safety, thermal engineering).

Jennifer Kurtz is a senior engineer at the National Renewable Energy Laboratory on the hydrogen technology validation team. As part of this team, Kurtz processes, analyzes, and reports on real-world data of fuel cell projects that span many fuel cell markets such as vehicles, forklifts, stationary, and backup power. Prior to joining NREL in 2007, she worked for six years at UTC Power, primarily in fuel cell system design and components. Kurtz received her master's degree in mechanical engineering from Georgia Tech and her bachelor's degree in physics from Wartburg College, Waverly, Iowa.

Robert G. Landers is currently an associate professor of mechanical engineering and the associate chair for graduate affairs in the Department of Mechanical and Aerospace Engineering at the Missouri University of Science and Technology. He received his PhD degree in mechanical engineering from the University of Michigan in 1997. His research and teaching interests are in the areas of modeling, analysis, monitoring, and control of manufacturing processes and alternative energy systems and has over 100 technical refereed publications. He received the Society of Manufacturing Engineers' M. Eugene Merchant Outstanding Young Manufacturing Engineer Award in 2004, is a member of ASEE and ASME, and is a senior member of IEEE and SME. Landers is currently an associate editor for the *IEEE Transactions on Control System Technology*, the *ASME Journal of Dynamic*

Systems, Measurement, and Control, and the *ASME Journal of Manufacturing Science and Engineering.*

Nima Lotfi is currently a PhD candidate in the Mechanical and Aerospace Engineering Department at the Missouri University of Science and Technology. He received his BSc in electrical engineering from Sahand University of Technology, Tabriz, Iran, in 2006 and his MSc in electrical engineering from Sharif University of Technology, Tehran, Iran, in 2010. Lotfi's research interests include nonlinear control and estimation design, modeling, and control of alternative energy systems including PEM fuel cells and Li-ion batteries to be employed in hybrid electric vehicles.

Kevin B. Martin is an assistant professor at Northern Illinois University in the Institute for the Study of the Environment, Sustainability, and Energy and the Department of Technology. He received his PhD from the Department of Engineering Management & Systems Engineering at the Missouri University of Science and Technology in 2009. Martin received a BS and an MS in chemical engineering from the University of Missouri-Rolla in 2002 and 2005, respectively. His primary research interests include the application of quantitative methodologies to investigate hydrogen supply chains. Martin served as team leader for the Missouri S&T EcoCAR team, which researched, designed, developed, and tested a full-size fuel cell plug-in hybrid electric vehicle. He has been involved in research projects sponsored by U.S. Federal Transit Administration, U.S. Department of Energy, U.S. Defense Logistics Agency, U.S. Department of Transportation–Research and Innovative Technology Administration, and U. S. Air Force Research Laboratory.

Andrew Meintz received his BS in electrical engineering from Missouri S&T in 2007. He continued his education at Missouri S&T toward a PhD as a U.S. Department of Education GAANN Fellow. His research interest is in power electronics, electrochemical energy storage systems, and hybrid electric vehicles. In 2008 and 2009 he interned at Sandia National Lab, first studying the effects of plug-in hybrid vehicles on the power grid and then on the stability effects of high photovoltaic penetration of an island grid. Beginning in 2009, he was involved with the Department of Energy–sponsored EcoCAR: The NeXt Challenge, a three-year collegiate competition to design and build fuel-efficient vehicles. As part of this challenge he designed, built, and tested a fuel cell plug-in hybrid vehicle. This led to his work on the vehicle, the energy management control design, and a vehicle simulator. Since completing his degree in December of 2011, Meintz has started his career with a position at General Motors as a high-voltage battery systems engineer.

Marc W. Melaina leads the hydrogen infrastructure analysis team at the U.S. Department of Energy's National Renewable Energy Laboratory (NREL). Before joining NREL in 2007, Melaina worked as a research track director at the Institute of Transportation Studies at the University of California at Davis. He has also worked for Argonne National Laboratory, the National Academy of Sciences, and the National Transportation Research Center at the Oak Ridge National Laboratory. Melaina completed his PhD through the School of Natural Resources and Environment at the University of Michigan, and has an MSE in civil engineering and BA in physics.

Vijay Mohan obtained his bachelor's degree in Mechanical Engineering in July 2008 from the P.E.S. Institute of Technology, which is affiliated with Visveswaraya Technological University in Belgaum, India. He joined the Missouri University of Science and Technology in Fall 2009 to pursue a master's of science in mechanical engineering. He worked under Dr. John W. Sheffield as part of the 2010 and 2011 Fuel Cell and Hydrogen Energy Association (FCHEA) Hydrogen Student Design Contests and also worked in the Missouri S&T's EcoCAR—The NeXt Challenge team. Mohan received his master's of science in December 2011.

Todd Ramsden specializes in the technical and economic analysis of fuel cell systems and supporting hydrogen infrastructures at the National Renewable Energy Laboratory (NREL) in Golden, Colorado. He currently focuses on the performance and market potential of fuel cell systems for automotive and material handling applications and life cycle assessments of energy use and greenhouse gas impacts of various potential hydrogen production and distribution pathways. Ramsden has extensive experience in the transportation sector, analyzing energy, greenhouse gas, and air pollution issues. He currently works in the Hydrogen Technologies and Systems Center at NREL. Prior to his work at NREL, he held positions at the U.S. Department of Transportation, Ford Motor Company, and the U.S. Environmental Protection Agency.

Carl H. Rivkin is the supervisor of the Codes and Standards Project team at the National Renewable Energy Laboratory (NREL) in Golden, Colorado. The Codes and Standards Project at NREL has responsibility for supporting the promulgation of Regulations, Codes, and Standards (RCS) required for the deployment of alternative fuels. Rivkin has over 25 years of experience in safety and environmental engineering including work at a regulatory agency. Prior to joining NREL, he worked for the National Fire Protection Association on alternative energy code projects. He was also the editor of *The NFPA Guide to Gas Safety*, published in 2005, which has several chapters devoted to hydrogen and flammable gas safety. Rivkin has a bachelor's degree in chemical engineering from the University of Michigan and an MBA from the University of Baltimore. He is a licensed Professional Engineer (PE) and Certified Safety Professional (CSP).

Steven F. Rodgers holds MS and BS degrees in mechanical engineering at the Missouri University of Science and Technology (Missouri S&T). His research focuses on modeling proton exchange membrane (PEM) fuel cells.

John W. Sheffield is professor of mechanical and aerospace engineering at the Missouri University of Science and Technology and associate director of the National University Transportation Center. Sheffield received a bachelor's degree in engineering science from the University of Texas at Austin, a master's degree in engineering mechanics from North Carolina State University, and a PhD in engineering science and mechanics from North Carolina State University. Over the years, he has held numerous positions at Missouri S&T. Sheffield has also served in a number of other research positions. More recently, Sheffield has served as an associate director of the United Nations Industrial Development Organization (UNIDO)–International Centre for Hydrogen Energy Technologies (ICHET) during 2005 and later as a consultant to UNIDO for ICHET Pilot Projects. Currently at Missouri S&T, Sheffield serves in key roles in the hydrogen energy technologies and advanced vehicle technologies projects funded by U. S. Air Force Research Laboratory, U. S. Defense Logistics Agency, U. S. Department of Defense, U. S. Department of Transportation–Research and Innovative Technology Administration, U. S. Federal Transit Administration, and U. S. Department of Energy.

Sam Sprik is a senior engineer in hydrogen technology validation at the National Renewable Energy Lab. As part of the Hydrogen Technologies and Systems Center, most of that time has been spent analyzing and developing software tools for the large amounts of data coming into the Tech Validation projects. Prior to hydrogen, he analyzed and built simulation code for hybrid electric vehicles at NREL in the Center for Transportation Technology and Systems. One of the tools that came out of that effort is named the ADVanced VehIcle SimulatOR or ADVISOR. While in graduate school at the University of Michigan, he worked in statistical quality control of manufacturing processes at Chrysler through a research assistantship.

Darlene Steward joined the staff of the National Renewable Energy Laboratory in January of 2007. She is a senior engineer in the Hydrogen Technologies and Systems Center primarily focusing on life cycle cost, energy, and systems analysis. She holds bachelor's and master's degrees in chemical engineering from the University of Colorado at Boulder. Her areas of expertise are environmental analysis and life cycle cost modeling. Her current research areas include cost and energy analysis of fuel cell–based combined heat and power systems using the fuel cell power model and analysis of hydrogen-based energy storage systems.

Lie Tang is a currently a postdoctoral researcher in the Department of Mechanical and Aerospace Engineering at the Missouri University of Science and Technology. He received his BE degree and ME degree in electrical engineering from Hohai University, Nanjing, China, in 2001 and 2005, respectively. He received his PhD degree in mechanical engineering from the Missouri University of Science and Technology in 2009. His research interests include modeling, simulation, and control of mechatronics systems, manufacturing processes, and hydrogen fuel cells.

Mathew Thomas is from Kottayam, Kerala, India. He has master of science degrees in mechanical engineering and engineering management from the Missouri University of Science and Technology, Rolla, Missouri. He has been an active participant in projects involving hydrogen, including the design, construction, and management of Missouri's first hydrogen fueling station. He was the team leader of the award-winning Hydrogen Education Foundation's Hydrogen Design Contest in 2008 and 2010. In 2009, he received the best paper award in hydrogen systems at the World Congress of Young Scientists on Hydrogen Energy Systems in Torino, Italy. He received his PhD in engineering management in May 2012.

Warren Santiago Vaz is a mechanical engineering doctoral student at the Missouri University of Science and Technology, from which he also holds BS and MS degrees in nuclear engineering. His research studies life cycle analysis and total cost of ownership of advanced vehicle powertrains.

Karen Webster has worked in oil refining, hydrogen analysis and engineering, transportation greenhouse gas reduction analysis, and is currently working primarily in biofuels and biochemicals. Webster holds a BS in chemical engineering from the University of California, Berkeley.

Keith Wipke is a senior engineer and manager of hydrogen analysis at the National Renewable Energy Laboratory, where he has worked in the area of advanced vehicles for nearly two decades. Wipke began his focus on hydrogen fuel cell vehicles in 2003 when he began leading NREL's participation in the Controlled Hydrogen Fleet and Infrastructure Demonstration and Validation Project (a.k.a. Learning Demo). NREL's Technology Validation team evaluates fuel cell technology in multiple applications, including cars, buses, forklifts, backup power, and stationary power. He also leads the Hydrogen Analysis Group, which includes hydrogen infrastructure analysis as well as technology validation staff. Wipke received his master's degree in mechanical engineering from Stanford University, is NREL's representative at the California Fuel Cell Partnership, and sits on the Board of Directors of the Fuel Cell and Hydrogen Energy Association (FCHEA).

Christopher Yang is a research engineer in the Sustainable Transportation Energy Pathways (STEPS) research program at the University of California–Davis Institute of Transportation Studies. He is co-leader of the infrastructure system analysis research group, and his research interests lie in understanding the role of advanced vehicles and fuels in helping to reduce transportation greenhouse gas emissions through infrastructure and system modeling. He works on hydrogen infrastructure systems, vehicle and electric grid interactions, and scenarios for long-term reductions in greenhouse gases from the transportation sector. Dr. Yang completed his PhD in mechanical engineering from Princeton University and a bachelor's and master's degree from Stanford University in environmental science and engineering.

Jarett Zuboy is a freelance researcher, analyst, and technical writer. He specializes in renewable energy topics and for 10 years has contributed to the National Renewable Energy Laboratory's hydrogen/fuel cell, solar, and transportation programs. He also has worked in the areas of utility energy efficiency, land and water conservation, and medicine. He holds a bachelor's degree in geology from Colorado State University.

Section I

Hydrogen Energy and Fuel Cell Modeling

1

Hydrogen and Electricity: Parallels, Interactions, and Convergence

Christopher Yang

CONTENTS

1.1 Introduction

The current energy system is comprised of a number of distinct energy carriers whose infrastructures have evolved over the course of the 20th century. The main fuels and energy carriers that consumers and end users use include petroleum fuels (gasoline and diesel), natural gas, and electricity. The transportation sector has been primarily powered by liquid petroleum fuels, while buildings and other end uses have relied on natural gas and electricity.

Recently, concerns about air pollution, oil and energy insecurity, and greenhouse gas emissions have been driving a search for cleaner energy sources and alternative energy carriers for all sectors, especially in the transportation sector. In particular, the last few decades have seen a renewed interest and significant research and development on electric drive vehicles, including battery electric vehicles, hybrid electric vehicles, and hydrogen fuel cell vehicles. Though there has been a great deal of activity in research and development, only a tiny fraction of our transportation energy use does not rely on petroleum.

While hydrogen has been touted primarily as a transportation fuel, it can serve a number of other potential needs and the potential development of a future hydrogen economy could significantly change the energy system because of linkages between hydrogen and the existing electricity system. The unique characteristics of electricity and its long history have resulted in an extensive infrastructure that converts primary energy resources such as fossil fuels, nuclear energy, and renewable energy resources into electricity and distributes the electricity to consumers essentially everywhere in the developed world. Any hydrogen infrastructure development can potentially take advantage of this expansive network of energy resource extraction and transport and electricity generation and distribution systems. The new energy system can also utilize the advantages of hydrogen to complement the use of electricity in some applications. And the development of a hydrogen energy system can take different forms depending upon how integrated a future one imagines for the co-evolution of the hydrogen and electricity systems.

1.1.1 Standard View of H$_2$ and Fuel Cells

Much of the interest and research in H$_2$ and fuel cells has been in the transportation sector, with many automotive companies developing low-temperature proton exchange membrane (PEM) fuel cell vehicle research, development, and demonstration (RD&D) programs in the last decade [1, 2]. Oil companies that primarily supply transportation fuels have also been involved with RD&D projects for H$_2$ production and refueling. Significant research and development is also being carried out on stationary fuel cells for use in the electric sector. However, hydrogen infrastructure is widely viewed as a transportation fuel supply system to be used in connection with fuel cell vehicles. Most stationary fuel cells do not require a ubiquitous hydrogen infrastructure since they are able to run on hydrocarbon fuels such as natural gas, which already has an extensive distribution infrastructure. Vehicles, on the other hand, require a widespread infrastructure to produce, store, transport, and dispense pure hydrogen at a network of refueling stations [3–6].

Because of the focus on hydrogen use in the light-duty transportation sector, the standard view of many in and out of the field is that hydrogen is a transportation fuel that will compete with and could potentially displace gasoline and diesel. Many hydrogen-related analyses and research programs focus primarily on hydrogen as a vehicle fuel [7–11]. Hydrogen and fuel cells are touted as an excellent alternative to gasoline and combustion vehicles because of their benefits with respect to efficiency, resource requirements, and environmental attributes [1, 8, 10–13]. The hydrogen infrastructure needed to extract, transport, and convert a primary energy feedstock to H_2 and store, transport, distribute, and dispense that hydrogen for use in personal vehicles is also analogous to the exploration, refining, distribution, and dispensing infrastructure for gasoline and diesel fuels. This focus can be thought of as an evolutionary model of H_2 and fuel cells because they are viewed as merely cleaner and more efficient technologies that will be used for light-duty vehicles. This framework is convenient because it does not fundamentally change the way that people view transportation fuels that power their vehicles. Hydrogen is merely a replacement for gasoline and fuel cells are a replacement for internal combustion engines.

1.1.2 Integrated View of H_2 and Electricity

In an alternative view, H_2 fuel and fuel cells are not merely replacements for specific components in the conventional transportation paradigm. Instead, they represent a new path that will be integrated with the electricity system, forming a future energy system with two primary energy carriers (hydrogen and electricity). There are multiple reasons for this convergence of hydrogen and electricity into an integrated hydrogen and electric energy system, including their complementary attributes as energy carriers, their potential production from the same primary energy resources, and their ability to be coproduced and interconverted.

H_2 and electricity are two decarbonized energy carriers that have very different yet complementary characteristics, which suggest specific uses and applications for each. With the emerging scientific, political, and public consensus on climate change, there will be an increasing impetus for reducing and eventually decarbonizing our energy system. Hydrogen and electricity are two energy carriers that enable conversion, transport, and utilization of a wide variety of primary energy resources in a decarbonized energy system.

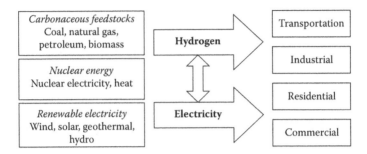

FIGURE 1.1
Schematic showing the parallel nature of hydrogen and electricity from the perspective of the energy resources and end-use sectors. (From Yang, C., *International Journal of Hydrogen Energy*, 33(8), 1977–1994, 2008.)

Another basic idea supporting the concept of hydrogen and electricity convergence is that hydrogen and electricity can and will be produced from the same primary energy resources and feedstocks, such as natural gas, coal, and biomass (see Figure 1.1). There are benefits associated with having another energy carrier, especially one that can be used in transportation applications that can be made from a large number of primary energy resources. However, this would also lead to a direct competition for the fossil, nuclear, and renewable energy resources that are used to produce each energy carrier.

The third argument for the convergence of hydrogen and electricity is related to the potential for their coproduction and interconversion. A number of studies have investigated production plants that can be used to generate both hydrogen and electricity [14–22]. In many of these studies, there are a number of benefits associated with producing both energy carriers in the same plant, including improved efficiency and lower costs. Interconversion is one of the most tangible examples of the shift toward a more integrated energy economy based upon hydrogen and electricity. With current energy carriers, there is little opportunity to convert between various forms. In addition, the widespread use and supporting infrastructure for these dual energy carriers may provide reliability benefits for consumers.

Figure 1.2 presents two different views of the hydrogen reactions in a fuel cell and electrolyzer. The "electrochemical" view shows the fuel cell reaction (on the right) that produces electricity when hydrogen and oxygen combine to form water and the electrolysis reaction (on the left) where electricity is required as an input to split water into hydrogen and oxygen. In this view, electricity and hydrogen have different roles: hydrogen is merely an enabler, while electricity is the primary focus (i.e., either the product or the input). This view is common when focusing on the end use of hydrogen—for example, if one thinks of a fuel cell vehicle as an electric vehicle that obtains its electricity from hydrogen.

The interconversion view describes the exact same reactions but emphasizes the conversions between energy carriers rather than the conversion between reactants and products of the electrochemical view. This alternative view

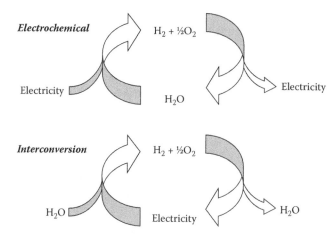

FIGURE 1.2
Alternative views of hydrogen and electricity reactions. The electrochemical view shows electricity as either an input or an output of chemical reactions, and the interconversion view shows water as an input or output of the conversion between H_2 and electricity. (From Yang, C., *International Journal of Hydrogen Energy*, 33(8), 1977–1994, 2008.)

shows that H_2 (plus O_2) and electricity are merely different forms of the same energy carrier that result from the addition and removal of water. Hydrogen is the hydrated form and electricity is the dehydrated form. This view emphasizes the large impacts that hydrogen production and conversion would have throughout the energy system, on the production, transmission, and conversion of energy. It is not the case that one view is better or worse than the other, but the significance of these two views is that they help to make clear, by emphasizing these different aspects, the relationship between H_2 and electricity.

This chapter will discuss many of the important elements that arise from the convergence between hydrogen and electricity as energy carriers, including possible opportunities and challenges. The goal is to help readers identify key areas of these future interactions and how they may impact the potential evolution of the future energy system.

1.2 Hydrogen and Electricity Parallels

Both hydrogen and electricity are energy carriers rather than energy sources, because they do not occur naturally but rather must be produced from other energy resources such as fossil fuels or renewables. A key similarity between hydrogen and electricity is that they are both zero-carbon and pollution-free energy carriers at the point of use and can have a wide range of *life cycle* emissions in bringing these energy carriers to the point-of-use.

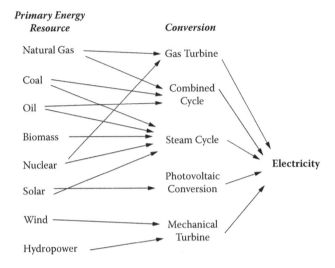

FIGURE 1.3
Resources and conversion technologies for electricity generation. (From Yang, C., *International Journal of Hydrogen Energy*, 33(8), 1977–1994, 2008.)

1.2.1 Generation Resources

As with electricity, hydrogen can be produced from a range of production methods and feedstocks. Figure 1.3 and Figure 1.4 show the potential resources for producing each energy carrier and their similarities. This is a major change, as hydrogen enables the possibility of using these resources

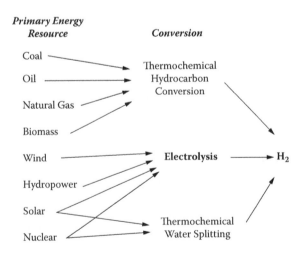

FIGURE 1.4
Resources and conversion technologies for hydrogen production. (From Yang, C., *International Journal of Hydrogen Energy*, 33(8), 1977–1994, 2008.)

in the transportation sector, which is currently, and has traditionally been, reliant on and restricted to petroleum.

Decarbonized, clean energy carriers that have multiple production pathways are valuable because they allow policies and resource constraints to affect the upstream side of the supply system without any inconvenience, or even knowledge of these changes, to consumers. Currently, a number of states have enacted a renewable portfolio standard (RPS), which mandates a specified fraction of electricity generation that must come from renewable resources, such as wind, solar, geothermal, and biomass. And while RPS targets are expected to increase over time, this process is transparent to the end user. Similarly, the ability to produce hydrogen from a wide range of resources enables producers, over time, to alter the mix of hydrogen production, so that it can be made less polluting and with lower greenhouse gas emissions as costs for these technologies declines. In fact, California has enacted a law (SB1505) that links state funding for hydrogen refueling stations to the renewable content and greenhouse gas emissions profile of the hydrogen that it dispenses (requiring a 30% reduction in greenhouse gas emissions and a goal of 33% renewable).

1.2.2 Generation Mix

Because of the variations in electricity demand that occur over the course of a day, and seasonally, and the difficulty in storing electricity, the electric power system consists of a number of power plants of different sizes and types, which are fueled by a number of energy sources. This structure has evolved because not all power plants need to be operating at full capacity all the time. Excess electricity generation that is not used cannot be stored efficiently and is thus wasted, so generation is carefully managed to make sure that there is the correct amount of generation occurring. Different types of power plants have different capital and operating costs associated with them, so some will be operated continuously while others will be operated only when demand is highest.

Hydrogen demand will also vary over the course of a year and the required output from a hydrogen production plant will not be constant over an entire year. While hydrogen can be stored more easily than electricity, it is not as inexpensive to store as a liquid fuel and hydrogen will likely not be stored for more than a few days. This means that variations in demand that occur on a longer timescale (i.e., seasonally) must be handled by the production plants themselves. Depending upon the extent of demand variation, it may be economically advantageous, like with the electric power system, for supply to consist of a mix of plants, with differing capital and operating costs, to minimize the cost of meeting demand.

1.2.3 Distribution and Infrastructure

Electricity is a commodity that is produced at hundreds or thousands of generating power plants in a given region and placed upon a common transmission and distribution infrastructure and then distributed nearly universally. Hydrogen infrastructure could also consist of a common network of hydrogen delivery that links a number of production facilities to the end users. Using H_2 primarily as transportation fuel would require delivery to a network of refueling stations spread throughout a region. If hydrogen were distributed to homes and businesses, an extensive network of pipelines would be needed that is similar to the network of natural gas distribution pipelines.

An analysis of regional hydrogen fuel infrastructure for supplying hydrogen FCVs in the U.S. states of Ohio and California indicates that because of economies of scale, having fewer large production plants provides lower cost fuel than many more smaller plants even if transportation distances are greater [6]. And while there are few production plants, they feed a common hydrogen delivery system that distributes hydrogen regionally to refueling stations in different cities.

1.3 Complementarity and Convergence

1.3.1 Complementary Attributes and Applications

H_2 and electricity are two decarbonized energy carriers that have very different yet complementary characteristics, which suggest specific uses and applications for each. Given the importance of reducing GHG emissions to avoid dangerous anthropogenic climate change, the use of decarbonized energy carriers is essential over the next few decades. Electricity is already in widespread use, and there is already a system for producing and distributing electricity from multiple sources to end users. Thus, switching to lower GHG-emitting sources of electricity can be done in a manner that is hidden from the end user with no disruption on their part.

The direct use of fuels for transportation and combustion applications (boilers, burners, and other applications) is currently optimized around the specific characteristics of the fuels they use (typically natural gas, gasoline, and diesel) and cannot generally be switched to lower- or zero-carbon fuels without upgrading to new technologies. There is the potential to use liquid biofuels to replace conventional fuels, but even in these situations, ethanol and biodiesel cannot replace gasoline and diesel without some modifications to current vehicles and engines. There also appears to be significant limitations in the amount of biofuels that can be sustainably produced [23, 24]. As a result, there are many benefits to the use of hydrogen in many of these applications that rely on direct use of fuels. These will be discussed in the context of the specific applications—vehicles and stationary applications.

1.3.1.1 Vehicles

Given the benefits associated with electric drive vehicles, hydrogen and electricity are in competition as the primary energy carrier for light-duty vehicles. It is necessary to have a decarbonized energy carrier on vehicles because it is too difficult to capture and collect CO_2 from dispersed mobile sources. And while biofuels may be able to provide a means to use conventional vehicle technologies in the light-duty vehicle sector, limitations in the amount of low-carbon biofuels may necessitate their use to other transportation sectors such as aviation, marine, and heavy-duty vehicles where it is more difficult to electrify [25, 26].

While a large number of automakers are actively researching and developing hydrogen fuel cell vehicles, many are also looking into battery electric vehicles (BEVs) and plug-in hybrid electric vehicles (PHEVs) [27, 28]. Battery electric vehicles have a long and interesting history that spans their introduction in the early 20th century and their brief resurgence and demise in the late 20th century [29]. Recently, a great deal of attention has been focused on the PHEV, which can be called, more specifically, a grid-connected, charge-depleting hybrid electric vehicle. Users connect these vehicles to the electricity grid to recharge the vehicle. The PHEV is flexible in that it runs on a combination of electricity and another fuel such as gasoline, so that it does not have the same refueling time and range limitations of a dedicated battery-powered vehicle [27, 28, 30].

Hydrogen is attractive for both vehicles and distributed generation because of its high conversion efficiency in a fuel cell. Unlike a gasoline engine, a fuel cell is typically more efficient at partial load—the typical state of a vehicle power plant—than at maximum power [27, 31, 32]. Improved vehicle efficiency, zero emissions, and the potential to use numerous domestic and renewable resources make hydrogen an attractive energy carrier for vehicles.

Electricity is, in many ways, the ideal alternative energy source for vehicles because of its numerous benefits: (1) electricity can be used very efficiently on board the vehicle, (2) it is quiet, (3) it has zero point-of-use emissions, (4) since electricity generation occurs at a few central locations, it is easier to regulate and control pollutant and greenhouse gas emissions, (5) it is much less expensive per mile driven than an equivalent amount of gasoline, and (6) it can be made from numerous domestic resources. The main challenges with electric vehicles are the cost and energy density of electricity storage on the vehicle, in the form of batteries [33]. There has been significant research and development into improving batteries, but still pure BEVs are currently unable to compete with conventional vehicles in terms of vehicle range, refueling time, and cost (related to the challenges of storing electricity). BEVs along with plug-in hybrid electric vehicles are currently commercial, and it remains to be seen if costs will decline to the point where they can move beyond niche vehicles and capture a large fraction of the mass market. Given the challenges with storage, electricity is generally better suited toward stationary applications, where electricity can be produced to match the demand, without the need for intermediate storage. In fact, fuel cell vehicles are electric vehicles with an alternative electricity storage system, which consists of hydrogen fuel and a hydrogen-to-electricity conversion device (the fuel cell). Fuel cell vehicles produce electricity on board the vehicle at the time and quantity that is needed to match the demands of the electric propulsion motor and vehicle auxiliaries, without the need for intermediate storage, although many are also hybridized and include an auxiliary battery. Yet fuel cells vehicles face their own technical, cost, and implementation challenges as well. One of the major questions for future alternative fuels and advanced vehicles is which system (batteries or hydrogen and fuel cells) will achieve the technical goals and cost reductions necessary to overcome the challenges with providing electricity for vehicle propulsion to make electric vehicles a commercial success. However, even this competition could be turned into a complementary relationship, by partitioning different vehicles depending upon their driving patterns. Given the energy storage limitations associated with batteries, some have suggested that battery vehicles can make sense for smaller low-power, limited-range vehicles (e.g., electric bikes, scooters, and neighborhood electric vehicles) and commuter vehicles [34]. These small BEVs would not compete directly with larger fuel cell vehicles but the vehicles would occupy different niches, which respond to different consumer needs. Another view of the complementary nature of hydrogen and electricity is an exciting technology that combines both into the same vehicle. Most of the demonstration fuel cell vehicles are hybridized and have an energy storage battery to enable regenerative braking and also to handle peak power events with a smaller fuel cell stack. Since PHEVs were a logical next step from hybrids, it is not a giant leap to think that the batteries in hybrid FCVs could be recharged by plugging in to create a fuel cell PHEV. This potential was highlighted when General Motors introduced their gasoline PHEV, the Chevy Volt, and also mentioned the future possibility of a fuel cell option for the PHEV. Some universities and research groups have also created FC-PHEV prototypes. This design would enable the vehicle to obtain electricity in two separate ways, from the grid via storage batteries as well as from a hydrogen fuel cell. This hybrid system would allow for the best utilization of each of the energy carriers attributes, including low cost for grid electricity (relative to fuel cell generated electricity) for short-distance driving around town and the greater energy storage and quick refueling of hydrogen relative to batteries for longer-distance trips.

These systems are also complementary in the sense that beyond the energy storage component of the vehicle, the remainder of the vehicle (including electric motors, motor controllers, power electronics, and electronic drive systems) are essentially the same and advances in one class of vehicle will spill over to improve the technology for other electric drive vehicles.

1.3.1.2 Stationary

While electricity is already in widespread use in residential, commercial, and industrial applications, there are several end uses in which it is not the primary energy carrier. Natural gas is the other major energy source/carrier used in a wide variety of applications and is directly combusted to provide heat for space heating, water heating, cooking, and industrial boilers and other process heat. Electrification is possible for many applications where fuel is traditionally used for heat. For example, heat pumps make it possible to use electricity in space- and water-heating applications at very high efficiency, compared with resistance heating.

However, there are a few opportunities where it may make sense for H_2 to be used in stationary applications. One is where the properties of hydrogen are similar to the properties of the fuel that is traditionally used and thus provides a potentially easier transition for the end user. An example of this is in industrial settings where high temperatures are needed and a burner or furnace that combusts hydrogen is similar to one that burns natural gas or other fuels. While natural gas is the most commonly used fuel for stationary applications, if a decarbonized energy carrier is needed to help reduce GHG emissions, hydrogen's similarities to natural gas may make the substitution easier to implement than electrification.

Another opportunity exists where there is potential for efficiency improvement, on the end-use side as well as on the production side. Combined heat and power (CHP) is a process in which a fuel is used to generate electricity at the point of use (i.e., distributed generation) but also where the waste heat in the process is of value. Because of the use of waste heat, which cannot be efficiently utilized at a power plant, a greater percentage of the energy in the original fuel becomes useful energy at the point of use. While CHP typically implies natural gas, it could be done with hydrogen as well. For example, hydrogen could be piped to large commercial or residential buildings, such as an office or apartment building, and a fuel cell would convert the hydrogen to electricity at around 50% efficiency. The remaining 50% of energy in the hydrogen would be converted to heat and most of this heat (~40%) could be used for heating water or space heating. In many circumstances, this combination would be more efficient than conventional electricity generation and distribution for meeting both electricity and heating needs for a building. There is also an additional upstream efficiency benefit associated with the production of H_2 relative to electricity. Hydrogen may be a more efficient energy carrier than electricity in some circumstances, because it can be made more efficiently than electricity from fossil fuels. It is more efficient to convert natural gas or coal into hydrogen than into electricity—typically there may be a gain of 20–25% in system efficiency. Thus, hydrogen may be preferable to electricity for integrated CHP applications from a resource usage and system efficiency perspective, though economic considerations will also play an important, perhaps the most important, role.

In a decarbonized energy future, hydrogen and electricity could be the two main energy carriers that can be used in stationary applications. Hydrogen is certainly not the best energy carrier for every end use, even some that rely predominately on natural gas today. On the other hand, electricity can meet the energy service demands for almost any end-use application, even heating. However, there are some instances where hydrogen can provide some advantages relative to electricity and it is important to analyze the economics and efficiency for each energy carrier and the entire energy pathway for use in a given end-use application.

1.3.2 Feedstock Competition

If one of the major areas for H_2 adoption is in the transportation sector, an important issue surrounding the shift away from petroleum-based transportation fuels is the convergence in primary energy feedstocks with electricity. One of the stated benefits of H_2 for use as a transportation fuel in the United States is to diversify energy resources and use domestic resources. Of course, these energy resources are the same ones that are used to make electricity.

Gasoline, which is predominantly used for light-duty vehicles, accounts for about 17.5% of the primary energy used in the United States, while electricity accounts for 40%. Given the efficiency improvements associated with FCVs on a life cycle basis, the primary energy used for 100% hydrogen-based light-duty transportation would account for between 20% and 30% of the primary energy for electricity generation. Growing use of personal vehicles and vehicle miles traveled or use of H_2 in other transportation sectors (i.e., heavy-duty) will likely increase required resources even more as would the use of hydrogen in nontransportation applications.

In the future, low carbon and renewable energy resources will become more constrained as awareness of the role of greenhouse gas emissions on climate change increases and measures and policies to reduce GHGs are enacted around the world. These changes will have a significant impact on the electric sector. Natural gas is one such feedstock that will play an important role in helping to reduce GHG emissions from a number of different sectors, including electricity and transportation. The total amount of electricity generated using natural gas has increased by 65% over the last decade [35, 36]. It is also one of the most likely near-term energy feedstocks for a developing hydrogen economy because of the technical maturity and cost of the steam reforming process. One of the key concerns in looking forward toward the convergence in energy feedstocks is what the impacts will be of these additional demands from a cost, supply availability, and environmental perspective.

Other important resources include biomass, nuclear, renewable electricity sources, and fossil energy with carbon capture and sequestration (CCS). Each of these resources is constrained in different ways—such as sustainable biomass resource limits, nuclear safety and public acceptance, intermittency and reliability challenges for renewable electricity, and safety and sequestration capacity for CCS—and thus will present challenges for widespread use for both electricity and hydrogen production.

1.3.3 Coproduction

Because hydrogen and electricity can be produced from the same feedstock resources, there are also a number of opportunities for utilizing these resources simultaneously in a facility to produce both energy carriers. This coproduction of hydrogen and electricity can potentially offer significant benefits for overall energy efficiency and economics including the following:[*]

1. Better heat and energy integration—multiple products and processes can allow for waste streams from one process to be utilized for the other, improving overall system efficiency.
2. Better scale economies—distinct demand for two separate products allows for redundant equipment to be eliminated and common equipment to benefit from scale economies.
3. Better equipment utilization—divergence in the timing of the demand for each of the products can allow for common equipment to be utilized at a higher rate.
4. Decarbonization benefits—technologies for reducing carbon can be applied to multiple products simultaneously.

This section will briefly discuss a number of coproduction strategies and studies that can be found in the literature [14–22, 37].

1.3.3.1 Large-Scale Thermo-Chemical Coproduction

Hydrocarbon fuels can be converted into hydrogen by high-temperature thermochemical processing such as partial oxidation (including gasification) and steam reforming. Other thermochemical methods of hydrogen production are also possible using high-quality, high-temperature heat (from nuclear or solar energy), though these methods are not yet mature technologies. These proposed methods attempt to take advantage of synergies in coproduction of electricity and hydrogen.

1.3.3.1.1 Syngas-Based Coproduction Options

The production of a synthesis gas (*syngas*), which is a mixture of H_2, CO, and CO_2, is a common industrial process that has been adapted for both hydrogen and electricity production. A number of studies have investigated the coproduction of electricity and hydrogen in a single fossil fuel plant [7, 14, 15,

---

[*] Not every coproduction plant will realize all of these benefits at all times and for all designs.

17, 19, 20, 37]. Hydrogen production from any hydrocarbon fuel such as coal, natural gas, or biomass would occur through the primary step of producing a syngas. Similarly, electricity production can be accomplished via syngas production. In the case of the integrated gasification combined cycle (IGCC) power plant, coal is gasified to produce the syngas, which is fed into a gas turbine generator and additional electricity is generated using a bottoming steam cycle. In a coproduction facility, these two production processes can be combined. Once the syngas is produced, hydrogen is separated out, and the remaining gas can still contain significant energy content in the form of CO and residual H_2. These remaining gases can be passed to a gas turbine or solid oxide fuel cell generator to generate electricity. These systems are compatible with carbon capture and sequestration (CCS) because the electric generator exhaust is predominantly CO_2 and water, which is easily removed. This CO_2-rich stream can be further purified, transported, and injected for storage into geologic formations such as depleted oil and gas reservoirs and saline aquifers.

In the early years of a transition to hydrogen, the demand for hydrogen may not be large enough to warrant building large central plants dedicated solely to hydrogen production. An important benefit of coproduction can be the utilization of "slipstream" hydrogen, which diverts a small stream of H_2 from an existing IGCC coal plant with carbon capture and sequestration. Additional equipment would be needed to capture hydrogen from this plant, and initial analyses of slipstream hydrogen has found that delivered hydrogen costs as low as $2/kg could be achieved at refueling stations that are near an existing IGCC with CCS [38]. This system could be cost-competitive with distributed natural gas steam reformers but have lower well-to-wheels (WTW) emissions. Coproduction is useful because combining H_2 production with electricity generation can help the plant to achieve large-scale economies to help lower the cost of H_2 production and provide flexibility in product output.

Additionally, the incremental cost of adding carbon capture and sequestration to hydrogen production is relatively small. Kreutz et al. show a smaller increase in price associated with the addition of carbon capture equipment on a hydrogen production plant (+14–19%) compared to the addition of carbon capture to electricity production (+32–36%) [20]. This is due to the fact that hydrogen production and separation produces a CO_2-rich stream regardless of whether CO_2 is vented or captured. As a result, the coproduction of hydrogen and electricity can help enable the cost-effective addition of carbon capture to electricity production, which could have an important impact on decarbonization efforts in the electric sector.

1.3.3.1.2 *High-Temperature Nuclear/Solar Cycles*
The use of these heat sources directly for H_2 production is currently only a detailed concept in the laboratory research phase and not a technology at even the pilot plant scale. The temperatures required for water splitting are

significantly higher than for conventional pressurized and boiling water reactors that make up the majority of nuclear power plants. Instead of electrolysis, splitting water to produce H_2 and O_2 can also be accomplished through a complex series of coupled chemical reactions driven by heat at temperatures between 400 and 1000°C from nuclear reactors or solar concentrators. A number of these thermochemical water-splitting cycles have been investigated for use with nuclear or solar heat and a recent assessment of these methods has identified the sulfur-iodine process as one of the most promising cycles [39].

General Atomics proposed a system based upon an advanced helium gas cooled reactor, Modular Helium Reactor (MHR) [39], designed to reduce the issues surrounding earlier nuclear reactors in terms of safety and efficiency. The system (H_2-MHR) is modular such that the helium reactor is physically separated from the hydrogen production plant via an intermediate heat loop. The intermediate heat loop can also be coupled with a gas turbine to produce electricity (GT-MHR). Coproduction, while not explicitly discussed in any of the studies, could be an excellent application for this technology, involving coupling of the intermediate heat loop with both a H_2 production facility and gas turbine or bottoming steam cycle. Given the modular nature of the system, coproduction using this technology could be quite flexible to vary H_2 to electricity output ratio depending upon the time of day or product revenue and provide benefits in terms of waste heat utilization, helium reactor utilization, and system flexibility.

Thermochemical water-splitting cycles are still undergoing research, are not as technically mature as fossil-based hydrogen production pathways such as steam reforming, coal gasification, or water electrolysis, and should be considered a longer-term possibility.

1.3.3.2 Small- to Medium-Scale Energy Stations

An energy station is a system that converts the energy in a feedstock such as natural gas into hydrogen for vehicle refueling, electricity, and possibly heat for use by the station or associated buildings. The three main parts of an energy station are (1) the hydrogen production unit, (2) the electricity generator, and (3) the H_2 compression, storage, and dispensing system. For some designs, system 1 and 2 can be integrated into one unit. Energy stations that also have an integrated cogeneration system that uses the waste heat from the electricity generator to help meet a building's heat and/or cooling loads are also referred to as combined heat, hydrogen, and power (CHHP) systems. By providing three value streams (vehicle fuel, building electricity, and building heating/cooling), CHHP systems potentially offer a faster return on the initial capital investment cost and the potential to lower H_2 costs relative to stand-alone distributed hydrogen stations.

Two fundamentally different energy station configurations have been proposed. The first is based upon a natural gas steam methane reformer (SMR) and a proton exchange membrane (PEM) fuel cell. The SMR is used for producing hydrogen, which can be compressed and stored for dispensing to H_2 fuel cell vehicles or diverted to a stationary fuel cell for the production of electricity and heat for use in stationary building applications [21, 22, 40, Li, 2011 #159]. The other configuration is based upon a high-temperature fuel cell for both hydrogen generation and electricity production. The high-temperature fuel cell (either solid-oxide [SOFC] or molten carbonate [MCFC]) can be fed natural gas and an internal reforming reaction occurs that creates a syngas that can be used for electricity production in the fuel cell. The system can be operated to alter the amount of energy from the anode syngas. Extracting more of the anode exhaust will allow for more hydrogen production because it is a mixture of H_2, CO, and CO_2 that can be purified, compressed, and stored for distribution to FCVs [16, 21, 41].

The benefits of these energy stations really come about when considering near-to-medium term hydrogen refueling stations. One of the major issues with these early stations is the low level of vehicle and hydrogen fuel demand, which reduces utilization of the station and its components and raises the cost of hydrogen produced by these systems. For the first energy station configuration, one of the major capital costs is the reformer and by coupling H_2 production with electricity generation, the energy station can increase the size and utilization of the reformer, which, due to economies of scale and higher capacity factor, can help lower the per unit hydrogen production costs. The system can also offer product flexibility so that, as hydrogen demand from vehicles grows over time, the system can shift the ratio of products to favor hydrogen production. By lowering near-term costs of hydrogen stations with low utilization, energy stations can help reduce some of the infrastructure hurdles associated with the challenging transition to a hydrogen economy. These energy stations are focused on the economics of the early hydrogen transition and do not appear to be viable long-term options because they are based upon natural gas resources without the ability to capture and sequester carbon.

1.3.4 Interconversion

Interconversion is a broad term that encompasses a wide range of potential interactions between hydrogen and electricity production. Interconversion describes the production of one energy carrier and subsequent transformation to the other energy carrier, including those that occur at different locations and scales. Interconversion is useful because it can allow for the production and transport of one energy carrier but the use of another energy carrier when its attributes are of particular value.

These systems can be classified into five different broad categories determined by the locations and order of production for each energy carrier:

1. Central electricity production, central H_2 production (intermittent renewables)
2. Central electricity production, distributed H_2 production (on-site electrolysis

3. Central H_2 production and central electricity production (fossil w/ CCS)

4. Central H_2 production and distributed electricity production (vehicle-to-grid or V2G)

5. Distributed H_2 and electricity production (E-station)

The main mechanisms for interconversion between H_2 and electricity are fuel cells that convert H_2 (and air) to electricity and electrolyzers that convert electricity into H_2 (and O_2). While a thorough review of all possible applications and options for interconversion is beyond the scope of this chapter, this section presents applications that may become common in the categories listed above.

1.3.4.1 Renewable Intermittent Electricity Storage

Hydrogen can be used to complement the production of renewable electricity, mainly as a means of energy storage, in cases where the electricity is generated intermittently and transmission is constrained. For example, large remote wind farms that generate electricity at low capacity factors (~30–40%) also underutilize transmission lines. The production and storage of hydrogen, coupled with a fuel cell, can generate electricity when the wind turbines are not, leveling the system electricity output and potentially sending more electrical energy over the same transmission lines [42]. Despite this possibility, there are challenges with this approach including the low utilization of the electrolyzer, which would be lower even than the capacity factor of the wind turbines. Proponents of this strategy suggest that hydrogen will be an important enabler for utilizing renewables in both the stationary and transportation sectors. Intermittent electricity can also be supplemented with grid-electricity to increase electrolyzer utilization and economics [43].

1.3.4.2 Off-Peak Electrolysis

One of the most often cited examples of interconversion is the electrolytic generation of H_2 at off-peak hours. In order for electrolysis to compete economically with fossil-based hydrogen production, low-cost electricity is essential, such as from off-peak coal or nuclear or excess renewable electricity. At 1–2 cents/kWh, $0.50–1.00 of the cost per kg H_2 comes from electricity, without even considering the cost of electrolyzers and other capital equipment, including transport, storage, and refueling. Studies have put the remaining (nonelectricity) costs between $0.70 and $4.00/kg for a refueling station that produces hydrogen onsite [7, 10]. Along with the price of electricity, these equipment costs and their utilization are major drivers in the cost of hydrogen via electrolysis. However, operation of electrolyzers during only off-peak hours (less than 1/3 of the day) leads to a trade-off between operating costs (in the form of lower average electricity costs) vs. capital costs (which would increase due to lower equipment utilization). Even with the low electricity costs, the low equipment utilization (for electrolyzers, compressors, and storage tanks) can lead to H_2 prices that may not be competitive with other H_2 sources.

1.3.4.3 Central H_2 Production and Electricity Generation (Power Plant w/CCS)

The production of hydrogen-rich syngas is an intermediate step in a number of industrial processes, including hydrogen production from coal, biomass, and natural gas, coproduction of hydrogen and electricity that was described in Section 1.3.3.1, and the generation of electricity in an integrated gasifier combined cycle plant (IGCC). The purification of hydrogen is generally only accomplished if H_2 is a desired end product or if decarbonization is required. Producing hydrogen in a fossil-fueled power plant will allow for separation of CO_2 from the energy-rich hydrogen stream, which can be combusted in a turbine or converted in a fuel cell to generate electricity. The production of H_2 (here as an intermediary rather than as a transportation fuel) enables precombustion carbon separation, which is an easier and less expensive means of carbon capture for storage purposes.

1.3.4.4 Central H_2 Production and Distributed Electricity Production (FCVs, MobileE, and V2G)

This category encompasses the energy pathways around using hydrogen in FCVs. Hydrogen will be produced centrally and then distributed to refueling stations and to vehicles, where it will be converted to electricity in a fuel cell stack for vehicle propulsion. Other important aspects include the evolution of "mobile energy," which is one of the innovative ideas surrounding the evolution in how consumers will interact with advanced electric drive vehicles. Fuel cell vehicles have the ability to produce clean electricity for use in a range of applications almost anywhere. The propulsive power requirements for these vehicles is on the order of 50–100kW, which is on the order of 50 times the average household power usage; thus, these vehicles have significant capabilities to bring a host of other activities to the vehicle setting that will help to provide additional value to electric drive vehicles relative to conventional vehicles [44].

If fuel cell vehicles were a substantial part of the fleet, their capacity could be sufficient to displace a significant portion of electricity use from the grid and also provide the grid with significant amounts of peaking power or ancillary services. The application of the hydrogen fuel cell power plant could be economical and provide significant benefits to the electricity grid because the vehicle and fuel cell engine capital cost is already sunk and electricity from stationary power plants providing ancillary grid services including spinning and regulation services can be quite expensive. However, significant obstacles to vehicle-to-grid (V2G) implementation exist, including issues relating to the utility permission, building the necessary grid-connection infrastructure, and adding the capability to the fuel cell vehicles [45].

1.3.4.5 Distributed H₂ and Electricity Production (E-Station and Building Systems)

A distributed model for H_2 and electricity production, the low-temperature energy station, is described in Section 1.3.3.2. In that system, a distributed natural gas reformer produces hydrogen for fuel cell vehicles and also electricity generation. The electricity production from the fuel cell would help serve the electrical needs of a building or location and would be complementary to the existing electricity grid and could enhance the reliability of its electricity supply. Another example of this model, without the H_2 vehicle refueling, is seen in the first commercial applications of fuel cells, for building stationary power. United Technologies Corporation (UTC) installed several hundred 200 kW phosphoric acid fuel cells (PAFC) in building applications. In these systems, natural gas is reformed to H_2 and then fed to the PAFC for electricity and heat cogeneration (combined heat and power), which can be very efficient (40% for electricity and 80% for heat and electricity) and very reliable. This is another nontransportation application of hydrogen that can lead to convergence and have important impacts on the stationary power sector.

1.4 Conclusions

In a future that is going to be increasingly carbon-constrained, the use of decarbonized energy carriers, hydrogen and electricity, will be critical to remove distributed GHG emissions from the system. A switch to H_2-based transportation could help in diversifying energy resource use for transportation as well as improve the environmental and climate footprint of this sector. The widespread use of H_2 for transportation as well as in stationary applications would open up new opportunities and challenges for integration with the rest of the energy system, especially the electricity sector. This chapter discusses some of the major ways that a future hydrogen economy would interact with the electricity sector and how the transportation and stationary fuels sectors and the electricity sectors might converge. H_2 and electricity are both zero-carbon, flexible, useful, and complementary energy carriers that could provide power for a wide range of applications. Hydrogen is touted as an important future transportation fuel in the light-duty sector because of its storage characteristics, efficiency, and emissions.

In addition, an important consideration for the evolution of the energy system is the competition and synergies for the use of energy resources for producing H_2 and electricity. Because of supply constraints for these resources, especially those that can help reduce greenhouse gas emissions, large additional demands could increase prices and affect the mix of resources used for electricity production. On the other hand, the option for coproduction and interconversion promote energy system efficiency, flexibility, and reliability because of the ability to couple the production processes and modify the outputs of each of the energy carriers. The complementarity between H_2 and electricity comes about because of their distinct characteristics that will be matched to the specific requirements of the application. Their convergence can also help promote sustainability because this hydrogen and electricity future enables zero pollution to the end user as well as a seamless and transparent transition from current production via fossil fuels to lower-carbon and renewable resources such as biomass, solar, and wind. A future hydrogen economy will not only be defined by the impact on the kinds of cars we drive and the way that we fuel those cars. It will likely be determined by the wide-ranging impacts (in terms of both opportunities and challenges) that this future hydrogen economy has on the other energy sectors, especially electricity, and how these two systems will interact and coevolve.

References

1. DOE. (2002). *National Hydrogen Energy Roadmap.* Department of Energy: Washington DC. p. 50.
2. DOE. (2005). *Hydrogen, Fuel Cells & Infrastructure Technologies Program: Multi-Year Research, Development and Demonstration Plan: Planned Program Activities for 2004–2015.* Department of Energy: Washington DC.
3. Ogden, J.M. (1999). Prospects for Building a Hydrogen Energy Infrastructure. *Annu. Rev. Energy Environ.* 24: 227–279.
4. Ogden, J.M., et al. (2005). *Technical and Economic Assessment of Transition Strategies Toward Widespread Use of Hydrogen as an Energy Carrier.* Institute of Transportation Studies, UC Davis: Davis, CA.
5. Yang, C., and J.M. Ogden. (2007). Determining the Lowest Cost Hydrogen Delivery Mode. *International Journal of Hydrogen Energy* 32(2): 268–286.
6. Johnson, N., C. Yang, and J.M. Ogden. (2008). A GIS-Based Assessment of Coal-Based Hydrogen Infrastructure Deployment in the State of Ohio. *International Journal of Hydrogen Energy* 33(20): 5287–5303.
7. H2A. (2005). *DOE Hydrogen Analysis Team (H2A).*
8. Greene, D., et al. (2006). Hydrogen Transition Modeling and Analysis HyTrans v 1.5, in *DOE Hydrogen Program Review.* Washington DC.

9. Ogden, J.M. (1999). Developing a Refueling Infrastructure for Hydrogen Vehicles: A Southern California Case Study. *International Journal of Hydrogen Energy* 24: 709–730.

10. NRC. (2004). *The Hydrogen Economy: Opportunities, Costs, Barriers, and R&D Needs*. National Research Council—Board on Energy and Environmental Systems: Washington DC. p. 394.

11. CalEPA. (2005). *California Hydrogen Blueprint Plan*. California EPA: Sacramento, CA.

12. Hoffmann, P. (2001). *Tomorrow's Energy*. Cambridge: MIT Press. p. 289.

13. Dunn, S. (2002). Hydrogen Futures: Toward a Sustainable Energy System. *International Journal of Hydrogen Energy* 27: 235–264.

14. Consonni, S., and F. Vigano. (2005). Decarbonized Hydrogen and Electricity from Natural Gas. *Energy* 30: 701–718.

15. Gray, D., and G. Tomlinson. (2002). *Hydrogen from Coal*. NETL, Editor, Mitretek.

16. Keenan, G. (2005). Validation of an Integrated System for a Hydrogen-Fueled Power Park, in *DOE Hydrogen Program Annual Review*. Washington DC.

17. Yamashita, K., and L. Barreto. (2005). Energyplexes for the 21st Century: Coal Gasification for Co-producing Hydrogen, Electricity and Liquid Fuels. *Energy* 30: 2453–2473.

18. Yang, C., and J.M. Ogden. (2007). *Co-production of H2 and Electricity*. UC Davis: Davis, CA. p. 22.

19. Chiesa, P., et al. (2005). Co-production of Hydrogen, Electricity and CO2 from Coal with Commercially Ready Technology, Part A: Performance and Emissions. *International Journal of Hydrogen Energy* 30: 747–767.

20. Kreutz, T.G., et al. (2005). Co-production of Hydrogen, Electricity and CO2 from Coal with Commercially Ready Technology. Part B: Economic Analysis. *International Journal of Hydrogen Energy* 30: 769–784.

21. Lipman, T.E., and C. Brooks. (2006). *Hydrogen Energy Stations: Poly-Production of Electricity, Hydrogen and Thermal Energy*. H.P. Program, Editor, Institute of Transportation Studies: Davis, CA. p. 32.

22. Lipman, T.E., J.L. Edwards, and D.M. Kammen. (2002). *Economic Analysis of Hydrogen Energy Station Concepts: Are "H2E-Stations" a Key Link to a Hydrogen Fuel Cell Vehicle Infrastructure?* University of California Energy Institute: Berkeley, CA. p. 81.

23. Perlack, R., et al. Biomass as a Feedstock for a Bioenergy and Bioprocessing Industry: The Technical Feasibility of a Billion-Ton Annual Supply, in *Joint Study sponsored by USDOE and USDA2005*, Oak Ridge National Laboratory.

24. Searchinger, T., R. Heimlich, R.A. Houghton, F. Dong, A. Elobeid, J. Fabiosa, S. Tokgoz, D. Hayes, and T. Yu. (2008). *Use of U.S. Croplands for Biofuels Increases Greenhouse Gases through Emissions from Land Use Change*. Sciencexpress.

25. Yang, C., et al. (2009). Meeting an 80% Reduction in Greenhouse Gas Emissions from Transportation by 2050: A Case Study in California. *Transportation Research Part D* 14: 147–156.

26. McCollum, D., and C. Yang. (2009). Achieving Deep Reductions in US Transport Greenhouse Gas Emissions: Scenario Analysis and Policy Implications. *Energy Policy* 37(12): 5580–5596.

27. Kromer, M.A., and J.B. Heywood. (2007). *Electric Powertrains: Opportunities and Challenges in the US Light-Duty Vehicle Fleet*. S.A. Laboratory, Editor, MIT: Cambridge, MA. p. 153.

28. Duvall, M., and E. Knipping. (2007). *Environmental Assessment of Plug-In Hybrid Electric Vehicles*. Electric Power Research Institute. p. 70.

29. Westbrook, M.H. (2001). *The Electric and Hybrid Electric Car*. Warrendale, PA: Society of Automotive Engineers. p. 198.

30. Markel, T., and A. Simpson. (2006). *Plug-In Hybrid Electric Vehicle Energy Storage System Design*. NREL: Golden, CO.

31. Srinivasan, S., et al. (1999). Fuel Cells: Reaching the Era of Clean and Efficient Power Generation in the Twenty-First Century. *Annu. Rev. Energy Environ* 24: 281–328.

32. Weiss, M.A., et al. (2000). *On The Road in 2020—A Life-Cycle Analysis of New Automobile Technologies*. MIT Energy Laboratory: Cambridge, MA.

33. NRC. (1998). *Effectiveness of the United States Advanced Battry Consortium as a Government-Industry Partnership*. National Academy of Science: Washington DC.

34. Sperling, D. (1995). *Future Drive: Electric Vehicles And Sustainable Transportation*. Island Press: Washington DC.

35. EIA. (2007). *Natural Gas Data*, available from: http://www.eia.doe.gov/oil_gas/natural_gas/info_glance/natural_gas.html [cited March 2, 2007].

36. EIA. (2007). *Annual Energy Outlook*. Energy Information Agency, US DOE.

37. Haeseldonckx, D., and W. D'Haeseleer. (2010). Using Renewables and the Co-production of Hydrogen and Electricity from CCS-Equipped IGCC Facilities, as a Stepping Stone towards the Early Development of a Hydrogen Economy. *International Journal of Hydrogen Energy* 35(3): 861–871.

38. Kreutz, T.G. (2005). The Potential of Hydrogen in a Climate-Constrained Future, in *AAAS Annual Meeting*. Washington DC.

39. Schultz, K.R., et al. (2003). Large-Scale Production of Hydrogen by Nuclear Energy for the Hydrogen Economy, in *National Hydrogen Association Annual Conference*. Washington DC.

40. Lipman, T.E., J.L. Edwards, and D.M. Kammen. (2004). Fuel Cell System Economics: Comparing the Cost of Generating Power with Stationary and Motor Vehicle PEM Fuel Cell Systems. *Energy Policy* 32: 101–125.

41. Margalef, P., et al. (2011). Efficiency of Poly-generating High Temperature Fuel Cells. *Journal of Power Sources* 196(4): 2055–2060.

42. Bartholomy, O. (2005). Renewable Hydrogen from Wind in California, in *National Hydrogen Association Annual Meeting*. Washington DC.

43. Saur, G., and T. Ramsden. (2011). *Wind Electrolysis: Hydrogen Cost Optimization*. National Renewable Energy Laboratory.

44. Williams, B.D., and K.S. Kurani. (2007). Commercializing Light-Duty Plug-in/ Plug-out Hydrogen-Fuel-Cell Vehicles: "Mobile Electricity" Technologies and Opportunities. *J. Power Sources* 166: 549–566.

45. Kempton, W., and J. Tomic. (2005). Vehicle-to-Grid Power Fundamentals: Calculating Capacity and Net Revenue. *Journal of Power Sources*. 144(1): 268–279.

46. Yang, C. (1979/2008). Hydrogen and Electricity: Parallels, Interactions, and Convergence. *International Journal of Hydrogen Energy*, 33(8): 1977–1994.

47. Yang, C. (1980/2008). Hydrogen and Electricity: Parallels, Interactions, and Convergence. *International Journal of Hydrogen Energy*, 33(8): 1977–1994.

48. Yang, C. (1982/2008). Hydrogen and Electricity: Parallels, Interactions, and Convergence. *International Journal of Hydrogen Energy*, 33(8): 1977–1994.

49. Yang, C. (1983/2008). Hydrogen and Electricity: Parallels, Interactions, and Convergence. *International Journal of Hydrogen Energy*, 33(8): 1977–1994.

2

Hydrogen Infrastructure: Production, Storage, and Transportation

Kevin B. Martin and Warren Vaz

CONTENTS

2.1 Introduction

The Energy Information Agency (EIA) predicts that global demand for energy will increase by 49% by 2035 (EIA, 2010). The projected increase in global energy demand compounded with economic repercussions from potential oil supply security issues and shortages has renewed interest in hydrogen as an energy carrier. Concerns about the ability to increase integration of renewable energy sources into the electric grid have also increased interest in the usage of hydrogen as an energy storage medium. The hydrogen production pathway can be optimized on a local and regional scale as hydrogen can be produced from any primary energy source. Hydrogen has additional benefits such as having a higher energetic production efficiency from synthesis gas than the production of Fischer-Tropsch synthetic fuels (Eucar, 2004).

This chapter provides a review of hydrogen production, storage, and transportation technologies. The production technologies selected represent promising near- and long-term options based upon state of technology, scale of production quantities, and environmental impacts. The production technologies include steam methane reformation, gasification, electrolysis, and thermochemical conversion. Compressed gas, liquid, cryo-compressed, metal hydride, and surface adsorption storage methods are presented based on the scale of hydrogen storage capability. The chapter concludes with a discussion on transportation methods and operational characteristics for an expanded hydrogen infrastructure.

2.2 Production

The selection of the appropriate hydrogen production process is based upon numerous factors including feedstock availability and cost, capacity and product purity requirements, and environmental concerns, among others. Several different technologies have been developed to utilize the variety of feedstocks including natural gas, coal, biomass, and water. Steam reformation of natural gas and other hydrocarbons takes place when the fuel reacts with steam at high temperatures (973–1373 K), producing hydrogen and carbon dioxide. In gasification processes, the fuel reacts with a controlled oxidant mixture (air or/and oxygen and stream), producing a combination of hydrogen and carbon dioxide. A third general production method involves decomposition of water via electrochemical and thermochemical cycles. In the electrolysis process, water is decomposed into hydrogen and oxygen via an electrochemical reaction. Thermochemical cycles involve extracting hydrogen from water at very high temperatures.

2.2.1 Steam Methane Reformation

Steam methane reformation involves four basic steps: desulfurization, reformation, carbon monoxide conversion, and purification. The first process involves natural gas, which is primarily methane (CH_4), being catalytically treated with hydrogen via a hydrogenation reaction to form hydrogen sulfide. The hydrogen sulfide is then removed via a scrubbing process in a ZnO bed following the reaction.

$$H_2S + ZnO \rightarrow ZnS + H_2O$$

Once the ZnO is saturated, air is used to regenerate the ZnO bed to form SO_2. Absorption is also widely used as a separation technique for removing hydrogen sulfide. In particular, the monoethanolamine (MEA), methyldiethanolamine (MDEA), and Purisol processes are commonly used. However, the low process operating temperatures limit their use to gas streams of light hydrocarbons.

When the feedstock contains high amounts of hydrocarbons greater than two carbon atoms a pre-reformer is utilized. The pre-reformer converts the higher carbon hydrocarbon chains into methane, carbon monoxide/dioxide, and hydrogen. Higher hydrocarbons are more reactive than methane and tend to form coke within the main reformer, thus deactivating the catalyst particles if not removed prior to being fed to the main reformer. The use of a pre-reformer also reduces the steam to carbon ratio in the reformer increasing the overall efficiency of the process and increasing the ability of the production plant to accept a variety of feedstock compositions.

The methane-rich stream is then fed to the main reformer where it is mixed with steam and passed over a catalyst such as nickel. Typically, the reaction is operated between 800 and 900°C with the steam being fed in a molar ratio of 2:1 to avoid coke formation. This step produces a synthesis gas, or syngas, which is composed of typically 75% hydrogen, 15% carbon monoxide, and 10% carbon dioxide.

$$CH_4 + H_2O \rightarrow CO + 3H_2$$

This is followed by a catalytic water-gas shift reaction to convert the carbon monoxide to hydrogen and carbon dioxide. The reaction is slightly exothermic with increases in temperature increasing reaction rates but resulting in the thermodynamic equilibrium favoring the left side of the reaction equation. Thus, a trade must take place between high flow rates with low conversion and low flow rates with high conversion. Thus, to achieve high flow rates and high carbon monoxide conversion percentages, a series of high and low temperature reactors are commonly used. This process is able to achieve 92% conversion of carbon monoxide into hydrogen (Spath and Mann, 2001).

$$CO + H_2O \rightarrow CO_2 + H_2$$

Finally, the hydrogen is purified, either using a pressure swing adsorption (PSA), gas separation unit (CO_2 scrubber), or membrane separator unit. Typically, modern production plants use the PSA process, which operates at 2 MPa and utilizes several molecular sieve adsorption beds in parallel to separate and purify the hydrogen gas. In a gas separation unit, the carbon dioxide

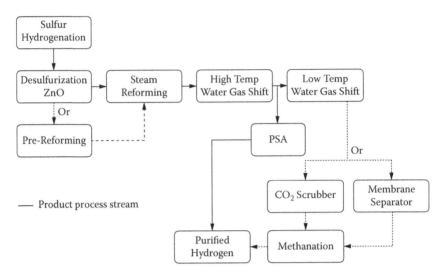

FIGURE 2.1
Large-scale steam methane hydrogen production.

is reacted with a solvent such as monoethanolamine, ammonia, methanol, water, or potassium carbonate, reducing the concentration of carbon dioxide to about 100 ppm. The carbon dioxide is further reduced in a methanation reactor by reacting it with hydrogen in the presence of a catalyst to produce methane. A membrane separator utilizes a selective membrane that separates the gas mixture based on molecular size. Figure 2.1 depicts a typical large-scale hydrogen production plant with all three types of purification.

Hydrogen production by steam methane reformation has several advantages. Steam methane reformation plants already exist and provide the majority of the hydrogen needed by industries in the United States and in the world. The overall efficiency of hydrogen production is between 40% and 60% depending on the specific process (Di Profio et al., 2009). This is one of the highest overall efficiencies of any hydrogen production process resulting in the lowest cost of any hydrogen production process. The feedstock, natural gas, is widely available and easy to handle. In addition, a vast natural gas distribution infrastructure already exists in the United States and in other parts of the world.

There are, however, some challenges. The hydrogen produced has to be purified before it can be used in fuel cells. Existing facilities are operating at maximum capacity, which means new facilities would have to be built or existing ones upgraded to meet any significant additional demand. The environmental impact is the greatest challenge since steam methane reformation results in the emission of carbon dioxide. To address the carbon dioxide emissions produced, many carbon capture and sequestration technologies have been developed. Large-scale reformation results in the emission of about 11 kg CO_2/kg H_2

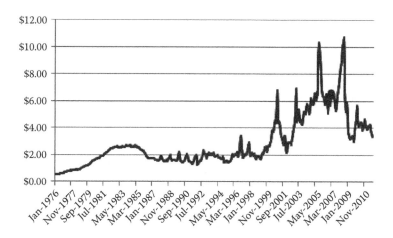

FIGURE 2.2
U.S. price of natural gas per thousand cubic feet at well head. (From the Energy Information Administration, U.S. Regional Natural Gas Prices, 2011.)

and small-scale reformation results in the emission of about 9 kg CO_2/kg H_2 (Lemus & Duart, 2010). Carbon sequestration can remove most of the CO_2 from the emissions (over 70%) (Lemus & Duart, 2010). An alternative to the direct capture and storage of carbon dioxide is the production of sodium carbonate, which is achieved by adding sodium hydroxide to the reformation reaction (Saxena et al., 2011). The addition of the sodium hydroxide also reduces the process temperature and alleviates the need for catalysts. However, the use of these technologies increases the cost of the hydrogen. In addition, the effects on the environment are not clear and may depend on how the carbon is stored. The volatile price of natural gas adds another issue to a complex problem. Figure 2.2 shows the monthly fluctuations in the price of natural gas in the United States. The price of natural gas is only expected to increase as usage for power generation increases and resources become increasingly difficult to exploit.

A typical hydrogen production rate when using the steam methane reformation process at petroleum refineries and chemical plants is 1.5 million m^3/day (122 tonnes/day) (Spath, Mann, 2001). However, most of this hydrogen is used as feedstock for various chemical processes. For comparison, current hydrogen fueling stations for vehicles have production rates of 2–200 kg/day (Fuel Cells, 2000, 2011), with Forsberg and Karlstrom (2007) determining that fueling stations with on-site steam methane reformation experienced optimal production rates at 1100–1400 kg/day.

2.2.2 Coal Gasification

Globally, there are approximately 200 years of recoverable coal reserves (BP, 2011). Coal is also a low-cost fuel; however, it contains a high ratio of carbon to hydrogen, resulting in significant carbon dioxide emissions. However, due to the stability in supply and price, coal will likely remain a significant feedstock in the energy supply portfolio into the future. To make a significant reduction in emissions, carbon dioxide generated from coal utilization needs to be captured and sequestered. A study by Smolinski et al. (2011) demonstrated that lignite coal produced slightly higher hydrogen production concentrations (66–67 vol%) than hard coal (59–64 vol%) or biomass (59–62 vol%) while utilizing a gasification process at 700°C. However, hard coal produced lower carbon dioxide emissions (18–23 vol%) than lignite coal (22–24 vol%) or biomass (30–32 vol%). Biomass produced the lowest carbon monoxide content (7–9 vol%).

A coal gasification system consists of coal preparation, gasifier, water gas shift reactor, gas separation unit, and a sulfur recovery unit. The coal is fed to the gasifier either as a slurry or dry. Within the gasifier, the coal is turned into syngas via the following reactions:

$$2C + O_2 \rightarrow 2CO$$

$$C + H_2O \rightarrow H_2 + CO$$

$$C + 2H_2 \rightarrow CH_4$$

In addition, the plant may utilize an air separation unit to feed pure oxygen to the gasifier. The use of oxygen in the gasifier increases the concentration of the carbon dioxide in the product stream from the gasifier. The higher concentration affords the opportunity to capture and sequester the carbon dioxide at a lower cost. Steam is also frequently employed as a gasification agent; however, carbon dioxide and hydrogen can also be used as gasifying agents. Prior to being fed to the water gas shift reactor, the syngas from the gasifier is cooled to remove ash particles. Once the solids are removed, the syngas is treated to remove sulfur before being fed to the water gas shift reactor. Within the water gas shift reactor, carbon monoxide produced within the gasifier reacts with water to form carbon dioxide and hydrogen. In a similar fashion to steam methane reforming, the product stream is then sent either to a pressure swing adsorption unit or to gas separation and methanation units for purification.

There are three main groups of gasifiers based upon fuel and oxidant flow directions: entrained flow (co-current flow), fluidized bed (countercurrent flow), and moving bed (countercurrent flow). The entrained flow gasifier exhibits carbon conversion of greater than 95%. The use of oxygen instead of air to entrain the pulverized coal results in avoiding the need to separate the nitrogen from product hydrogen later in the process. A key point concerning this process is the ability of this technology to accept all types of coal as a feedstock. However, as the coal must be pulverized into a fine powder, this type of gasifier is not suitable for use with biomass unless it undergoes pyrolysis beforehand.

In a fluidized bed gasifier, coal is consistently fed either from above the bed or directly into the bed where it is mixed with partially gasified and fully gasified coal. The fluidizing medium is usually air, although oxygen or steam may also be used. The bed of coal is fluidized by flowing gas from the bottom of the gasifier, causing the pulverized coal particles to float within the bed. As the particles are gasified, the particles become light enough to be entrained out of the reactor. It is also important to maintain the temperature within the bed below the initial ash fusion temperature of the coal to avoid particle agglomeration. Fluidized bed gasifiers are most useful for fuels that form highly corrosive ash, such as biomass fuels, which damage the walls of slagging gasifiers.

In a moving bed gasifier, coal is fed from the top of the reactor with steam, oxygen, or air fed from the bottom of the reactor. Unlike in a fluidized bed reactor, the pulverized coal collects and creates a solid bed of particles. The by-product ash is collected and removed from the bottom of the reactor. Moving bed gasifiers typically have high methane content in the produced gas. Moving bed gasifiers do produce tars and oils that contain sulfur and ammonia, which complicate the gas cleanup. The complex composition of the product gas results in large capital investments to handle by-products, relatively high maintenance costs, and relatively small throughput compared to land usage.

Hydrogen production from coal is a promising near- to mid-term technology. The development of integrated gasification combined cycle (IGCC) power plants provide a large-scale, dual production (electricity and hydrogen) capability pathway that could address early uncertainty with hydrogen demand. Although early IGCC plants experienced very high capital cost, next-generation IGCC plants are expected to have simplified designs reducing capital costs and higher thermal efficiencies. This is in part by using oxygen and carbon dioxide to gasify the coal. Figure 2.3 shows a process layout for an IGCC plant.

2.2.3 Electrolysis

During an electrolysis process, water undergoes disassociation into hydrogen and oxygen when an electric current is applied. Electrolyzers may be classified as either unipolar or bipolar. Unipolar electrolyzers are simply systems that consist of a single tank in which the electrodes are separated by a membrane or diaphragm. As the electrodes are connected in a parallel fashion, unipolar electrolyzers operate under high-current, low-voltage conditions. Bipolar electrolyzers contain individual electrodes separated by insulators that allow one side of a single electrode to act as cathode while the opposite side serves as the anode for the adjacent electrolysis cell. The electrodes in

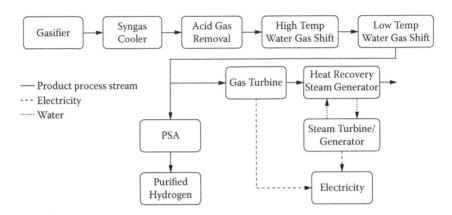

FIGURE 2.3
Integrated gasification combined cycle (IGCC) plant.

a bipolar electrolyzer are wired in series resulting in a low-current, high-voltage device.

Electrolyzers are also classified based upon production technology. Currently, there are three main types of electrolyzers under development: alkaline, polymer electrolyte membrane, and solid oxide electrolyzer cells. Alkaline electrolyzers typically use a 30 wt% aqueous solution of potassium hydroxide as the electrolyte. A potassium hydroxide solution is used as it has high conductivity and the energy loss related to the oxygen evolution reaction is minimized. This solution is fed into the electrolyzer along with the water feedstock. The operating temperature of an alkaline electrolyzer is typically limited to 100°C to limit alkali corrosion of electrodes. However, there has been research at operating at significantly higher temperatures (400°C) (Ganley, 2009). Alkaline electrolyzers typically use iron, nickel, and nickel alloy as materials for the electrodes and can be either unipolar or bipolar. A bipolar alkaline electrolyzer may have more than 100 cells packaged together in a filter press arrangement to form a stack with each cell fed water in parallel. The electrolyzer has the following reaction kinetics:

Cathode

$$2H^+ + 2e^- \rightarrow H_2$$

Anode

$$2OH^- \rightarrow \frac{1}{2} O_2 + H_2O + 2e^-$$

Overall

$$H_2O \rightarrow \frac{1}{2} O_2 + H_2$$

The second type of unit is a proton exchange membrane (PEM) electrolyzer. Unlike in an alkaline electrolyzer, a solid ion conducting membrane is used to separate the electrodes and transport protons from the anode to the cathode. In this unit, the electrolyte is a solid ion conducting membrane and the electrolyzer is fed only pure water. This membrane has a polymer structure with sulfonic acid groups attached that allow for proton transfer. Electrodes are typically loaded with catalysts such as platinum and ruthenium. The efficiency of a PEM electrolyzer increases as the operating temperature increases. This is achieved as the chemical reaction rates at the electrodes are increased, thus reducing the overpotential and amount of electrical energy required. However, the membrane must stay hydrated to function properly, which typically limits the operating temperature to below 100°C. Current research on high-temperature PEM electrolyzers that operate around 130°C is taking place (Xu et al., 2011). PEM electrolyzers have several advantages over alkaline electrolyzers such as higher purity of product gases, lower power consumption, less ecological impact, and smaller overall dimensions of systems. However, water purity requirements are much higher for PEM electrolyzers.

Solid oxide electrolyzer cells (SOECs) have gained much attention to generate hydrogen from electricity as they operate at high temperatures (approximately 500–950°C), which increases the reaction kinetics and reduces electrical energy requirements (Jin et al., 2011). SOECs use a solid ceramic material as the electrolyte that selectively transfers negatively charged oxygen ions to the anode. Unlike PEM and alkaline electrolyzers, water is disassociated into hydrogen and oxygen ions at the cathode. Nuclear power systems and renewable energy sources, as well as waste heat from high-temperature industrial processes, could provide the electricity and heat required for operation of SOECs (Stoots et al., 2010). SOECs can also operate in autothermal mode; this operation does not require a high-temperature source for the electrolysis reaction but only one capable of vaporizing sufficient water for the desired production rate. (See Figure 2.4.)

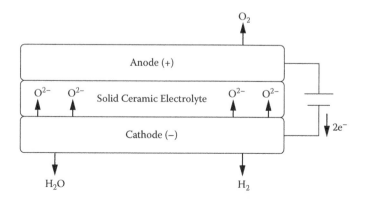

FIGURE 2.4
Solid oxide electrolyzer cell.

In general, electrolyzers provide a carbon-free pathway for hydrogen production when renewable energy is utilized. Recently, there has been a focus on a wind-hydrogen systems that utilize electrolyzers (Martin and Grasman, 2009; Carton and Olabi, 2010). In addition, they are capable of being scaled to a variety of production rates leading to higher utilization rates. However, challenges remain including reducing the capital costs of the equipment, improving the energy efficiency of the electrolyzer, and producing larger systems if a centralized production system is selected. A method that could lead to higher overall system efficiencies is operating the electrolyzer at higher pressures required for storage, therefore lowering or avoiding separate compressor costs.

2.2.4 High-Temperature Electrolysis and Thermochemical Conversion from Nuclear Energy

Hydrogen produced by nuclear power plants is very pure and suitable for use in hydrogen fuel cells, which require up to 99.999% purity. A 600 MW nuclear reactor would be able to produce 640000 m^3/day of hydrogen along with 175 MW of power (Naterer et al., 2008). Hydrogen can be produced utilizing nuclear power by utilizing either solid oxide electrolyzer cells (SOECs) and/or thermochemical conversion. Using generation IV reactors creates additional benefits as they will have passive safety features, increased proliferation resistance, improved economics, and better management of radioactive waste. However, generation IV reactors are still only in the design phase and are not expected to be in operation before 2020. In particular, the very high-temperature reactor (VHTR) is a generation IV reactor type currently being developed due in part to its potential for hydrogen production.

High-temperature electrolysis using nuclear energy is currently being developed by a number of countries including Canada, France, and South Korea (Ryland et al., 2007; Yvon et al., 2009; Koh et al., 2010). Helium is the primary coolant and is used to drive a gas turbine to provide the electricity needed for electrolysis. See Figure 2.5. Some of the heat contained within the primary coolant is also transferred using a heat exchanger to the water before electrolysis. The addition of this high-temperature process heat decreases the amount of energy that must be converted from heat (from the nuclear reactor) to electricity (from the generator) and then to chemical energy (in the hydrogen molecule), thereby raising the overall efficiency of the process. With operating temperatures up to 1000°C, the VHTR can be coupled with certain thermochemical processes to produce hydrogen with an efficiency of around 50% (Naterer et al., 2008).

Several thermochemical processes for hydrogen production have been developed based on the very high-temperature reactor (Vilim et al., 2004). Oak Ridge National Laboratory is currently pursuing the sulfur-iodine cycle for hydrogen production. Canada is developing the copper-chlorine cycle for hydrogen production with the goal of using it with future high-temperature

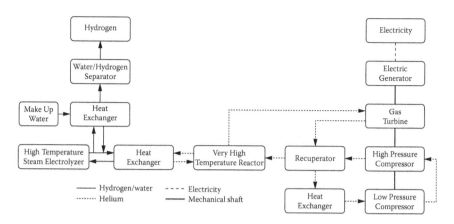

FIGURE 2.5
VHTR with steam electrolysis plant.

gas-cooled reactors. It is estimated that the thermochemical splitting of water to produce hydrogen using the heat from generation IV nuclear power plants will be able to provide hydrogen at between 4.3 and 6.8 ¢/kWh by 2020 (Lemus & Duart, 2010). For comparison, it is estimated that for a 600 MW thermal capacity, a direct Brayton cycle power plant with an outlet temperature of 900°C, the cost of hydrogen will be about 9.69 ¢/kWh (O'Brien et al., 2010).

The copper chlorine cycle involves the following equations:

$$2\ CuCl + 2\ HCl \rightarrow 2\ CuCl_2 + H_2 \qquad (<100°C)$$

$$2\ CuCl_2 + H_2O \rightarrow CuO \cdot CuCl_2 + 2\ HCl \qquad (400°C)$$

$$2\ CuO \cdot CuCl_2 \rightarrow 4\ CuCl + O_2 \qquad (500°C)$$

This cycle has the ability to use low-grade waste heat and requires a relatively low voltage. It requires temperatures only up to 500°C, whereas certain other processes require temperatures up to 800°C. In addition, the reactants are relatively common and react to completion without side reactions (Naterer et al., 2010).

The S-I cycle involves the following equations:

$$I_2 + SO_2 + 2\ H_2O \rightarrow 2\ HI + H_2SO_4 \qquad (120°C)$$

$$2\ HI \rightarrow I_2 + H_2 \qquad (450°C)$$

$$2\ H_2SO_4 \rightarrow 2\ SO_2 + 2\ H_2O + O_2 \qquad (830°C)$$

This cycle is a purely thermal process, which is suitable for large-scale hydrogen production. It has a high thermal efficiency of up to 50% and is an all-fluid process, which makes it easier to expand and to achieve continuous operation (Zhang et al., 2010).

There are some challenges that must be overcome in order for nuclear energy to provide the energy needed to produce large-scale quantities of hydrogen. The main drawback is that nuclear power plants are highly capital intensive. In the United States, the construction of nuclear power plants generally takes four years from ground breaking to operation. This does not include the time needed by the Nuclear Regulatory Commission to license the plant, which can be up to two years and precedes any construction. The initial investment can range from $5 billion to $14 billion (based on budgets for nuclear power plants built after 2005) (WNA, 2011). The long-term storage of nuclear waste must also be addressed if nuclear energy is to become the main source of hydrogen production. The Nuclear Waste Policy Act of 1982 specifies that spent nuclear fuel is to be disposed of in deep geological repositories. However, no such repository is currently in operation in the United States and nuclear utilities are forced to store their spent nuclear fuel on-site.

2.2.5 By-Product and Industrial Hydrogen

Hydrogen is an important raw material in the chemical, petroleum, metallurgical, pharmaceutical, electronic, and food industries with a total of 18 MM tonnes produced in 2010 (Garvey, 2011). During the production of many industrial processes, hydrogen is also produced as a by-product. The total generation of by-product hydrogen is significant in the United States as 3.4 MM tonnes were produced in 2006 (EIA, 2011). The largest consumers of hydrogen in the United States are oil refineries (4.08 MM tonnes), ammonia synthesis facilities (2.62 MM tonnes), and methanol production plants (0.39 MM tonnes) with a total of 7 MM tonnes consumed in 2003 (Suresh et al., 2004).

Although these industries are consumers of hydrogen, they may also serve as important locations of hydrogen supply during the introduction of hydrogen as an energy carrier. For example, petroleum refineries and commercial hydrogen production facilities could serve as a hydrogen source in California and Texas. Approximately 275 vehicles/day could be fueled by using only 1% of the hydrogen produced at a typical petroleum refinery. In the Midwest, ammonia production facilities could serve as a hydrogen source for the region as it is produced as a feedstock. Petroleum refineries in New Jersey and Pennsylvania could serve the Northeast with hydrogen. Another important process is the chloralkali process, which is used to produce chlorine, sodium hydroxide, and other chemicals. Approximately 70% of the U.S. chloralkali production facilities are in the Gulf region (Thornton, 2000). This industry produces about 0.35 MM tonnes of hydrogen every year (EIA, 2011). The cost of recovered by-product hydrogen depends on the merchant, but it is generally less than the cost of hydrogen that is produced at a dedicated facility.

2.3 Storage and Transportation

Regardless of whether production is centralized or decentralized, the hydrogen that is produced will require some form of storage. The choice of storage would depend on the type of application, the energy density needed, the quantity to be stored, the storage period, the capital and operating costs, local resources, and the safety of the method. Each of the several storage methods has its own advantages and challenges. They can be classified as large-scale or small-scale, chemical or physical, onboard or stationary.

2.3.1 Large-Scale Storage

To support a fully developed hydrogen economy, large-scale storage facilities may be required to maintain adequate reserves between production facilities and small-scale storage facilities.

2.3.1.1 Cryogenic

Cryogenic hydrogen has a density nearly twice that of compressed hydrogen at 70 MPa. Liquid hydrogen is stored in specially insulated cryogenic tanks under pressure, which have provisions for cooling, heating, and venting. Tank sizes can range from 1.5 m³ (100 kg) to 75.0 m³ (5000 kg). However, liquefaction is an energy-intensive process. Estimates are about 12.5–15.0 kWh/kg for liquefaction compared to about 6.0 kWh/kg for compression to 70 MPa (Di Profio et al., 2009). Due to the energy-intensive nature of liquefaction, the CO_2/kg of hydrogen stored is the highest for cryogenic storage (Di Profio et al., 2009).

Currently, there are approximately 450 large-scale liquid hydrogen storage sites in the United States spread across 41 states (EIA, 2011). One of the largest tanks for storing liquid hydrogen is located at Cape Canaveral, Florida, and has a capacity of 3800 m³ (245 tonnes). If cryogenic storage of hydrogen is adopted, new storage facilities would have to be developed. The initial capital investment for a new storage facility is high due to the need for liquefaction equipment as well as storage vessels. The operating costs are also high due to the energy-intensive liquefaction process. The cost and the energy needed to liquefy hydrogen decreases per kilogram of hydrogen liquefied as the plant capacity increases. Typical liquefaction capacities can range from 100 kg/h to 10000 kg/h and typical on-site storage capacities can range from 115000 kg to

900000 kg (Amos, 1998). In the United States, the total liquefaction capacity is approximately 69000 tonnes/day (EIA, 2011). Cryogenic storage of hydrogen is economical at large production rates and at long storage periods.

2.3.1.2 Underground

Natural underground formations such as aquifers and depleted natural gas and man-made caverns can be used to store hydrogen. Aquifers are an underground layer of water-bearing permeable rock or sand trapped between layers of impermeable rock. Hydrogen can be compressed and injected to form a gas pocket that is contained underground by water below and by impermeable rock above it. Hydrogen can also be stored in the porous rock found in natural gas fields in a similar fashion as that used in aquifers. The storage of gases underground has taken place since the early 1900s when natural gas began being stored in depleted oil fields. Man-made caverns such as salt caverns are created by pumping fresh water into a salt dome and dissolving the salt. The resulting cavern can then be used to store, among other things, hydrogen. The pressure depends on the amount of gas in the cavern and decreases as the gas is depleted. A surface lake connected to the bottom of the cavern can maintain a constant pressure equal to the head of the lake, as long as the gas and the rock do not dissolve in water. Underground mines with suitable geological features can be used in a similar way.

Currently, there are only two underground hydrogen storage sites in the United States. One of these facilities is Clemens Terminal in Texas with a capacity of approximately 30 million m^3. In comparison, there are approximately 400 underground storage sites utilized for natural gas with a total capacity of 102 billion m^3 (EIA, 2011). However in 2001, the total capacity for underground storage of natural gas in the United States was 238 billion m^3 (Forsberg, 2005). Therefore, there is a potential to modify and utilize some existing natural gas storage for large-scale hydrogen storage. The utility of many of these sites is also increased as they are located close to industrial and urban centers.

Countries such as France, Germany, Russia, and the United Kingdom have experience storing hydrogen as a mixture of gases in aquifers and salt caverns without gas losses and minimal safety concerns. Once a site has been developed, most of the costs associated with its operation are from hydrogen compression, which is independent of the actual storage. Properly characterizing the geology of the site to ensure a stable facility for storage is the challenge in the development of new sites. Based on natural gas storage experience, an underground facility capable of storing 238,000 tonnes of hydrogen would cost between $200 million and $400 million or between $0.80/kg and $1.60/kg of hydrogen (Forsberg, 2005). A more recent estimate places the cost at about $1.20/kg, which makes underground storage the most economical option among currently available storage options (Amos, 1998; Lemus,

Duart, 2010). However, storing hydrogen on such a large scale may not be needed for an extended period.

2.3.2 Small-Scale Storage

A hydrogen economy will require small-scale storage to distribute and utilize hydrogen in various applications. Small-scale storage is characterized by relatively small amounts of hydrogen and relatively short storage periods. Vehicles that run on hydrogen, whether fuel cell or internal combustion, may require hydrogen storage at a fueling station in addition to onboard storage. Residential and portable power applications such as home fueling, stationary fuel cells, and emergency backup power units may also require small hydrogen storage capabilities.

2.3.2.1 Compressed

Compressed hydrogen must be stored in specially designed tanks capable of withstanding the storage pressures, which can range from 17 MPa to 70 MPa. These tanks are usually made of steel. However, tanks made of carbon fiber lined with aluminum, steel, or specific polymers are used when weight is considered. When compressed, the density of hydrogen at 35.0 MPa is about 23 kg/m^3 and at 70.0 MPa is about 38 kg/m^3. This leads to an energy density of 767 kWh/m^3 (27°C, 35 MPa).

The volume of the storage tank is the biggest challenge, since the density of compressed hydrogen is lower than that of liquid hydrogen. Compression of hydrogen is an energy-intensive process that increases the overall cost. Estimates are about 6.0 kWh/kg for compression to 70 MPa, which leads to the CO_2/kg of hydrogen stored to be high (approximately 1.3 kg of CO_2/kg of hydrogen) (Di Profio et al., 2009). However, compression consumes only a third of the energy that liquefaction does. In addition to the cost of compressing hydrogen, the cost of compressed storage tanks must also be taken into account. The cyclic loading of tanks, which tend to heat up as they are filled with compressed hydrogen, reduces tank life.

Compressed storage is mostly done aboveground. However, underground storage is also possible and of particular interest for fueling stations as it decreases the amount of land used. This reduces the chance of accidents since the storage tanks are isolated, but increases the difficulty associated with inspection and maintenance. There are approximately 600 small-scale compressed hydrogen storage sites in the United States (EIA, 2011). Storage capacities range from 100 kg to 1300 kg and storage pressures range from 1 MPa to 30 MPa (Amos, 1998). Hydrogen fueling stations typically have storage capacities that range from 10 kg to 150 kg and storage pressures range from 1 MPa to 45 MPa with up to 100 MPa required in the future to simultaneously fast fill several 70 MPa vehicles.

2.3.2.2 Cryogenic and Cryo-Compressed Hydrogen

Stationary cryogenic storage is mostly done aboveground, although underground storage is possible. Hydrogen vehicles in particular benefit from the greater energy density when compared to compressed hydrogen. Although boil off of hydrogen has been an issue in the past, recent advances in tank design show that venting takes place only during periods of infrequent use (Michel et al., 2006). The energy required to liquefy hydrogen consumes 30% of the chemical energy stored based upon the lower heating value (LHV) of 120 MJ per kg of hydrogen. Due to this and several other factors, including losses during transfer and packaging of these systems into vehicles, many automakers have elected to focus on other storage methods.

The cryo-compressed tank is an alternative storage that combines using high pressure at cryogenic temperatures. The method is flexible in fueling source, using liquid, gas, or cooled gas. Hydrogen stored as a supercritical fluid can exceed the density of liquid hydrogen resulting in vessels capable of containing 2–3 times more fuel than conventional compressed hydrogen vessels (Aceves et al., 2010).

2.3.2.3 Metal Hydride

Metal hydrides are compounds of one or more metal cathodes (M^+) and one or more hydride anions (H^-). When pressurized, most metals bind strongly with hydrogen, resulting in stable metal hydrides that can be used to store hydrogen conveniently on board vehicles. Examples of metal hydrides are $LaNi_5H_6$, MgH_2, and $NaAlH_4$. Metal hydrides can be liquids or powders that are usually stored in tanks at approximately 1 MPa. As the pressure is reduced or the temperature is increased (between 120°C and 200°C), hydrogen is released. The metal hydride can be recharged without the use of a high-pressure compressed gas or cryogenic liquid. When designing efficient metal hydride systems, the critical material properties to manipulate are thermal conductivity, heat of reaction, and activation energy (Jorgensen, 2011).

Metal hydride storage has a low risk of accidental leaks since the hydrogen is stored within the metal hydride crystal and requires energy to be released. In addition, the energy spent in storing hydrogen using metal hydrides is about half as much as compression (70 MPa) and about a fifth as much as that of liquefaction. Thus, the CO_2/kg of hydrogen is the lowest of any storage method due to the low storage and release energies involved. Metal hydride storage also has an energy density (kWh/m^3) that is about three times higher than compressed and cryogenic storage (Di Profio et al., 2009).

However, there are some challenges as well. Metal hydride storage has a relatively high energy density by volume (kWh/m^3) but a relatively low energy density by weight (kWh/kg). Typical values are 1% to 9% by weight (Zhang et al., 2010). This results in metal hydride storage tanks that are about four times heavier (250–300 kg) than gasoline tanks. Even with improvements in metal hydride technology, storage tanks are currently too heavy for use in passenger vehicles and are limited for use in other applications. The cost of metals such as lanthanum and lithium raises the overall cost of the storage systems to a level that limits its adoption to small-scale applications. Early systems exhibited slow uptake rates, further limiting appropriate applications. Concerning safety, metal hydrides tend to react violently with moist air, making handling a challenge. Hydrogen purity is also a concern as impurities are detrimental to the performance of metal hydrides. The efficiency and reversibility of metal hydride storage systems over a large number of cycles and the thermal management of these systems are also proving to be challenging.

2.3.2.4 Surface Adsorption

A relatively recent proposition for hydrogen storage is using surface adsorption. At cryogenic temperatures, hydrogen is effectively adsorbed on to porous materials such as activated carbon and carbon nanotubes. As the temperature approaches room temperature, the weight density of stored hydrogen decreases. However, as the pressure increases, so does the weight density of stored hydrogen. To release hydrogen, some energy must be supplied. While the energy density by weight is similar to that of metal hydrides, the energy density by volume is six times lower. Weight densities of 5.5% to 19% have been reported (Assfour et al., 2011). In addition, a well-to-wheel analysis yielded a cost of 20.3 ¢/km for carbon nanotube storage (de Wit, Faaij, 2007).

2.3.3 Transportation

There are three main methods available to transport hydrogen: (1) trucks, trains, and barges carrying compressed or liquefied hydrogen; (2) pipelines transporting compressed hydrogen; and (3) chemical carriers such as hydrocarbons and hydrides that are reacted at the point of use. Gaseous tube trailer and cryogenic liquid truck delivery play an important part in the distribution of hydrogen. Tube trailers are capable of delivering 300 kg at 18 MPa. Cryogenic tanks capable of being delivered by semi-truck operate at 1.1 MPa and are capable of transporting 3200–4500 kg of hydrogen. Studies have shown that pipeline transportation is the most cost-effective and energy-efficient method to transport hydrogen over long distances and large quantities (Leighty et al., 2006; Keith and Leighty, 2002). Currently, there are 1100 kilometers of pipeline transportation in the United States. Hydrogen pipelines typically operate at pressures around 1–2 MPa, but operating pressures may reach 10 MPa (IEA, 2005). Sodium borohydride ($NaBH_4$) and methanol (CH_3OH) are leading chemical hydrogen carriers. These carriers could be used for mass transportation of hydrogen with the hydrogen released either before distribution or by the final customer. Sodium borohydride can be stored under ambient conditions as a stable aqueous solution without any safety precautions besides alkalinity concerns. The hydrogen release process entails bringing the solution into contact with a catalyst. The sodium borohydride system does have the advantage that the carrier is liquid; however, the decomposition product is slurry of $NaBO_2$, which has to be recycled. Methanol is another potential chemical carrier of hydrogen. Methanol is a liquid at normal operating temperatures and pressures and can be handled in much the same way as gasoline. In general, this process entails central production of methanol, which is then transported by conventional methods to end users where it is reformed into hydrogen. Methanol is typically produced from natural gas, although Li et al. (2010) describes a coal-based methanol production and distribution for hydrogen in China.

2.4 Conclusions

A key driver toward energy independence is the cost reduction of hydrogen. This is particularly evident concerning the adoption of hydrogen as a transportation fuel. In order for hydrogen to be adopted for transportation use, it must be comparable to conventional fuels and technologies on a per-kilometer basis. In support of this ultimate goal, many concepts such as combined heat and power (CHP) (Braun et al., 2006) systems and combined heat, hydrogen, and power (CHHP) (Becker et al., 2011) systems have shown great promise at reducing the cost of hydrogen technologies. Hydrogen can be produced using diverse, domestic resources including natural gas and coal (with carbon sequestration), biomass, nuclear, and renewable energy technologies, such as wind, solar, geothermal, and hydroelectric power. The diversity of feedstock lends itself to investigating whether selected production technologies may become regional.

A second aspect of a future hydrogen economy, which is partially dependent on the production technology, is the selection of a distribution method. Some technologies such coal gasification, thermochemical processes, and high-temperature electrolysis lends them to a large-scale, centralized production method. Steam methane reformation and electrolysis processes are scalable within a range allowing these technologies to operate either in a decentralized or centralized mode. Centralized production units typically are more energy efficient and if greenhouse gases are by-products would afford the potential opportunity to sequester them. However, large-scale, efficient transportation methods would have to be developed to minimize additional transportation costs. In addition, centralized production plants would require large initial capital investments in a developing market. A decentralized approach would mean producing the hydrogen locally or on-site instead of a centralized facility, thereby alleviating the need for large-scale transport of hydrogen. Decentralized production also allows for a gradual investment in hydrogen infrastructure. However, small-scale hydrogen generation typically has a higher cost and if natural gas reformation is utilized would nearly alleviate any carbon capture possibility due to the high costs. In summary, the mass acceptance of hydrogen, including the infrastructure selection, is complex with multipurpose and specialty applications showing the greatest opportunities for near-term deployment and success.

References

Aceves, S., Espinosa-Loza, F., Ledesma-Orozco, E., Ross, T., Weisberg, A., Brunner, T., & Kircher, O. (2010). High-denisty automotive hydrogen storage with cryogenic capable pressure vessels. *International Journal of Hydrogen Energy* 35, 3, 1219–1226.

Assfour, B., Leoni, S., Seifert, G., & Baburin, I. A. (2011). Packings of carbon nanotubes—New materials for hydrogen storage. *Advanced Materials* 23, 10, 1237–1241.

Becker, W. L., Braun, R. J., Penev, M., & Melaina, M. (2011). Design and technoeconomic performance analysis of a 1 MW solid oxide fuel cell polygeneration system for combined production of heat, hydrogen and power. *Journal of Power Sources*, 200, 2, 34–44.

Braun, R. J., Klein, S. A. & Reindl, D. T. (2006). Evaluation of system configurations for solid oxide fuel cell-based micro-combined heat and power generators in residential applications. *Journal of Power Sources* 158, 2, 2006.

British Petroleum. (2011). Statistical review of world energy. Retrieved from http://www.bp.com/productlanding.do?categoryId=6929&contentId= 7044622.

Carton, J. G., & Olabi, A. G. (2010). Wind/hydrogen hybrid systems: Opportunity for Ireland's wind resource to provide sustainable energy supply. *Energy* 35, 12, 4536–4544.

de Wit, M. P., & Faaij, A. P. C. (2007). Impact of hydrogen onboard storage technologies on the performance of hydrogen fuelled vehicles: A techno-economic well-to-wheel assessment. *International Journal of Hydrogen Energy* 32, 18, 4859–4870.

Di Profio, P., Arca, S., Rossi, F., & Filipponi, M. (2009). Comparison of hydrogen hydrates with existing hydrogen storage technologies: Energetic and economic evaluations. *International Journal of Hydrogen Energy* 34, 22, 9173–9180.

Energy Information Administration. (2010). Annual Energy Outlook 2010. Retrieved from http://www.eia.gov/oiaf/archive/aeo10/index.html.

Energy Information Administration. (2011). U.S. Regional Natural Gas Prices. Retrieved from http://www.eia.gov/emeu/steo/pub/cf_tables/steotables. cfm?tableNumber=16.

Energy Information Administration. (2011) Electric Power Annual. Retrieved from http://www.eia.gov/cneaf/electricity/epa/epa_sum.html.

Energy Information Administration. (2011). Natural Gas Monthly. Retrieved from http://www.eia.doe.gov/oil_gas/natural_gas/data_publications/natural_ gas_ monthly/ngm.html.

Energy Information Administration. (2011). The Impact of Increased Use of Hydrogen on Petroleum Consumption and Carbon Dioxide Emissions. Retrieved from http://www.eia.doe.gov/oiaf/servicerpt/hydro/index.html.

Eucar, C. (2004). Well-to-wheels analysis of future automotive fuels and power trains in the European context. Joint Research Centre of the European Commission.

Forsberg, C. (2005). Futures of hydrogen produced using nuclear energy. *Progress in Nuclear Energy* 47, 1–4, 484–495.

Forsberg, P., & Karlström, M. (2007). On optimal investment strategies for a hydrogen refueling station. *International Journal of Hydrogen Energy* 32, (5): 647–660.

Fuel Cells 2000. (2011). Hydrogen Fueling Stations in the U.S. Retrieved from http://www.fuelcells.org/info/charts/h2fuelingstations-US.pdf.

Ganley, J. (2009). High temperature and pressure alkaline electrolysis. *International Journal of Hydrogen Energy* 34, 9, 3604–3611.

Garvey, M. (2011). The Hydrogen Report After Losing Some Momentum in 2010, Hydrogen Markets Pick Up. CryoGas International.

International Energy Agency. (2005). Prospects for Hydrogen and Fuel Cells. IEA Technolgy Analysis Series.

Jin, C., Yang, C., & Chen, F. (2011). Characteristics of hydrogen electrode in high temperature steam electrolysis process. *Journal of the Electrochemical Society* 158, 10, B1217–1223.

Jorgensen, S. (2011). Hydrogen storage tanks for vehicles: Recent progress and current status. *Current Opinion in Solid State and Materials Science* 15, 2, 39–43.

Keith, G., & Leighty, W. (2002). Transmitting 4,000 MW of New Windpower from North Dakota to Chicago: New HVDC Electric Lines or Hydrogen Pipeline. Retrieved from http://www.leightyfoundation.org/files/ND-Chicago-HVDC-H2pipeline.pdf.

Koh, J., Yoon, D., & Oh, C. H. (2010). Simple electrolyzer model development for high-temperature electrolysis system analysis using solid oxide electrolysis cell. *Journal of Nuclear Science and Technology* 47, 7, 599–607.

Leighty, W. C., Holloway, J., Merer, R., Somerday, B., San Marchi, C., Keith, G., & White, D. E. (2006). Compressorless hydrogen transmission pipelines deliver large-scale stranded renewable energy at competitive cost. *International Gas Union World Gas Conference Papers*, 4, 1787–1811.

Lemus, R. G., & Martínez Duart, J. M. (2010). Updated hydrogen production costs and parities for conventional and renewable technologies. *International Journal of Hydrogen Energy* 35, 9, 3929–3936.

Martin, K.B., & Grasman, S. E. (2009). An assessment of wind-hydrogen system for light duty vehicles. *International Journal of Hydrogen Energy* 34, 16, 6581–6588.

Michel, F., Fieseler, H., & Allidieres, L.(2006). Liquid hydrogen technologies for mobile use. *World Hydrogen Energy Conference*, Vol. 16.

Naterer, G. F., Fowler, M., Cotton, J., & Gabriel, K. (2008). Synergistic roles of off-peak electrolysis and thermochemical production of hydrogen from nuclear energy in Canada. *International Journal of Hydrogen Energy* 33, 23, 6849–6857.

Naterer, G. F., Suppiah, S., Stolberg, L., Lewis, M., Wang, Z., Daggupati, V., ... Avsec, J. (2010). Canada's program on nuclear hydrogen production and the thermochemical Cu-Cl cycle. *International Journal of Hydrogen Energy* 35, 20, 10905–10926.

O'Brien, J. E., McKellar, M. G., Harvego, E. A., & Stoots, C. M. (2010). High-temperature electrolysis for large-scale hydrogen and syngas production from nuclear energy—Summary of system simulation and economic analyses. *International Journal of Hydrogen Energy* 35, 10, 4808–4819.

Ryland, D. K., Li, H., & Sadhankar, R. R. (2007). Electrolytic hydrogen generation using CANDU nuclear reactors. *International Journal of Energy Research* 31, 12, 1142–1155.

Saxena, S., Kumar, S., & Drozd, V. (2011). A modified steam-methane-reformation reaction for hydrogen production. *International Journal of Hydrogen Energy* 36, 7, 4366–4369.

Smoliński, A., Howaniec, N., & Stańczyk, K. (2011). A comparative experimental study of biomass, lignite and hard coal steam gasification. *Renewable Energy* 36, 6, 1836–1842.

Spath, P., & Mann, M. (2001). *Life Cycle Assessment of Hydrogen Production via Natural Gas Steam Reforming* (NREL Publication No. TP-570-27637). Golden, CO.

Stoots, C., O'Brien, J., Condie, K., & Hartvigsen, J. (2010). High-temperature electrolysis for large-scale hydrogen production from nuclear energy—Experimental investigations. *International Journal of Hydrogen Energy* 35, 10, 4861–4870.

Suresh, B., Schlag S., & Inogucji, Y. (2004). *Chemical Economics Handbook Marketing Research Report*. SRI Consulting.

Thornton, J. (2000). *Pandora's Poison: Chlorine, Health, and a New Environmental Strategy*. MIT Press.

Vilim, R. B., Feldman, E. E., Pointer, W. D., & Wei, T. Y. C. (2004). Generation IV Nuclear Energy System Initiative—Initial VHTR Accident Scenario Classification: Models and Data. Argonne, IL: Nuclear Engineering Division, Argonne National Laboratory.

World Nuclear Association. (2011). Nuclear Power in France. Retrieved from http://www.world-nuclear.org/info/inf02.html.

World Nuclear Association. (2011). Nuclear Power in the USA. Retrieved from http://www.world-nuclear.org/info/inf41.html.

World Nuclear Association. (2011). The Economics of Nuclear Power. Retrieved from http://www.world-nuclear.org/info/inf02.html.

Xu, W., Scott, K., & Basu, S. (2011). Performance of high temperature polymer electrolyte membrane water electrolyser. *Journal of Power Sources* 196, 21, 8918–8924.

Yvon, P., Hittner, D., & Delbecq, J. M. (2009). Perspectives for the French R&D program for high and very high temperature reactors. *2008 Proceedings of the 4th International Topical Meeting on High Temperature Reactor Technology*, HTR 2008, 1, 67–72.

Zhang, P., Chen, S. Z., Wang, L. J., Yao, T. Y., & Xu, J. M. (2010). Study on a lab-scale hydrogen production by closed cycle thermo-chemical iodine-sulfur process. *International Journal of Hydrogen Energy* 35, 19, 10166–10172.

3

PEM Fuel Cell Basics and Computational Modeling

Umit O. Koylu, Steven F. Rodgers, and Scott E. Grasman

CONTENTS

3.1 Introduction

Fuel cells directly produce power through the electrochemical reaction of a fuel and oxygen. A typical setup places the fuel and oxidizer streams on opposite sides of a semipermeable barrier (Figure 3.1). In the most popular type of fuel cell, Proton Exchange Membrane or Polymer Electrolyte Membrane (PEM) fuel cells, this barrier is, as the names imply, a polymer electrolyte that is permeable to protons (H^+ atoms). It is, however, impermeable to electrons. This means that electrons are compelled to travel around the membrane through wires, creating an electrical current that can be utilized.

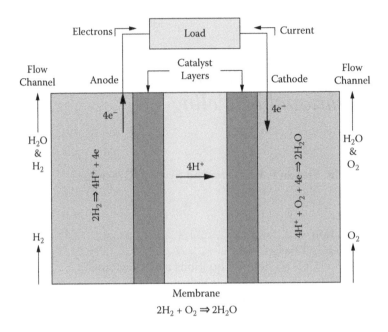

FIGURE 3.1
Proton exchange membrane (PEM) fuel cell.

Despite the seeming simplicity of this process, the details of a fuel cell's operational mechanism and the incurred losses are very complex and not fully understood (Biyikoglu, 2005). To gain insight into the fundamental processes occurring in fuel cells, many different types of models have been developed and published. Some of these are derived from first principles and some by incorporating empirical correlations with working fuel cells. Models are also of practical interest to industry, as they allow a designer to predict the behavior of fuel cells to reduce the time and expense of prototyping and testing. For industrial models, ease of use, ease of incorporation into a larger system, and an appropriate balance of comprehensiveness and computational cost are more important than a complete characterization of all aspects of the fuel cell.

A practical model of industrial interest has been developed by Gamma Technologies® (GT®) (Wahiduzzaman et al., 2004). Since the GT® software package is already in use for the simulation of vehicle systems and internal combustion engines (Vudumu and Koylu, 2009a), the inclusion of a PEM model could help facilitate the incorporation of fuel cells into transportation applications. Before this can happen, however, it must be determined if the model output is consistent with working fuel cells. The model also omits the effects of humidity on a fuel cell, which is an aspect of fuel cell function that is of great concern in fuel cell systems (Buchi & Srinivasan, 1997). A functional model of humidity dynamics should then be developed for incorporation with the current fuel cell model to increase its scope and applicability to designers of fuel cell systems.

Much of the work in modeling PEM fuel cells has been derived from the models of Springer et al. (1991) and Bernardi and Verbrugge (1992). The Bernardi and Verbrugge model was one-dimensional, steady state, and isothermal, assumed a fully hydrated membrane, and incorporated the Nernst-Planck, Schlögl's velocity, Butler-Volmer, and Stefan-Maxwell equations. The Springer et al. (1991) model was similar but did not use porous-electrode equations, and it accounted for the effects of membrane (specifically Nafion 117) water content on membrane water diffusion, electro-osmotic drag, and membrane conductivity. Membrane water content was based on its relationship to the water activity in the fuel cell as described by Zawodzinski et al. (1991).

These models were later extended by Fuller and Newman (1993) and Nguyen and White (1993) who considered flow along the channels and heat and mass transfer effects. These models, as well as those they were derived from, were valid only in the absence of liquid water. Two-phase flow was considered by Wang et al. (2001), who developed a model to predict liquid water formation and its effect on electrochemical kinetics and transport at the cathode. Further advances were made by Murgia et al. (2002), who eliminated the nonlinear portions of the Bernardi and Verbrugge model, yielding a model that was more stable and less computationally intensive. Pisani et al. (2002a) corrected for inaccuracies of the Bernardi and Verbrugge model at high current densities by taking into account the effects of water flooding in the cathode. Pisani et al. (2002b) also worked to replace as many of the fitting coefficients as possible with mechanistic derived coefficients.

Dutta et al. (2000) developed one of the first three-dimensional models by using the commercially available software package Fluent® to solve the complete three-dimensional Navier-Stokes equations in the flow channels. Berning et al. (2002) later developed a model using the program CFX®, incorporating the effects of all major transport phenomena except water phase change. Other three-dimensional, finite element models have been made using CFD-ACE+® (Mazumder and Cole, 2003) and Star-CD® (Meng and Wang, 2004a; 2004b).

While multidimensional models are necessary to understand the details of fuel cells, they are nevertheless computationally demanding. Simpler quasi-dimensional models are desirable for fast computations with reasonable accuracy for practical design, control, and optimization purposes. Such fuel cell simulations also provide cost-effective technical tools that considerably shorten the development time from conceptual ideas to actual products. This is especially important for PEM fuel cell technologies that are in the initial stages of development and commercialization. Hence, there is a crucial need to develop, validate, and utilize simple yet predictive models for PEM fuel cells.

Wahiduzzaman et al. (2004) presented a PEM fuel cell model for the software package GT-Suite®. Although they demonstrated the model and obtained reasonable predictions, they did not validate it against experimental data. Originally, this software has been developed to simulate internal combustion engines and has become an industry standard for many automotive companies. It has been widely used in the literature for predicting the performance of conventional gasoline and diesel engines. Very recently, its use has also been extended to successfully compute combustion and emission characteristics of hydrogen-powered engines. While this model is not as comprehensive in its formulation and omits multidimensional effects, its implementation in an integrated computational package allows all relevant subsystems to be simulated in the same environment, which could be attractive to industries wishing to incorporate fuel cells into designs. This is especially important for vehicles where fuel cells can power electric motors or be combined with advanced batteries.

Based on the above observations, the objectives of this chapter are first to introduce the basics of PEM fuel cell operations and then to discuss a simple model that can be used as a cost-effective technical tool for fast computations with reasonable accuracy so that practical design, control, and optimization of PEM fuel cells can be accomplished for portable and stationary industrial and military applications. The fuel cell model and its implementation in a software package that is already being widely used to simulate internal combustion engines by many automotive companies will especially be useful for hybrid vehicles in the transportation industry. In particular, a model capable of incorporating the effects of humidity on fuel cell performance will be developed and implemented, and the new model will also be compared with data from fuel cells available in the literature. The model's accuracy and suitability to computationally predict the operational performance of PEM fuel cells will also be evaluated. The results presented here are expected to contribute to the improved design and analysis of PEM fuel cells and therefore lead to a faster and smoother transition to emerging cleaner and more efficient energy conversion devices in the power industry.

3.2 PEM Fuel Cell Basics

3.2.1 Fuel Cell Modeling Equations and Constants

3.2.1.1 Reversible Cell Voltage

The electrochemical reaction that drives a hydrogen fuel cell can be expressed as Equation 3.1 (anode reaction) and Equation 3.2 (cathode reaction), or in a combined form as Equation 3.3.

$$H_2 \rightarrow 2H^+ + 2e^- \tag{3.1}$$

$$H^+ + 2e^- + \tfrac{1}{2}O_2 \rightarrow H_2O \tag{3.2}$$

$$H_2 + \tfrac{1}{2}O_2 \rightarrow H_2O \tag{3.3}$$

The electrical potential of a fuel cell reaction is given by

$$\Delta \hat{g}_f = -nFE^o \qquad (3.4)$$

where $\Delta \hat{g}_f$ is the Gibbs free energy of the reaction, n is the number of electrons transported, v_i is the Faraday constant, and E^o is the maximum theoretical electrical potential, or voltage. This ideal voltage is affected by the gas pressure and concentration such that actual maximum reaction voltage is given by the Nernst equation:

$$E = E^o + \frac{RT}{nF} \ln \frac{\Pi a_{reactants}^{v_i}}{\Pi a_{products}^{v_i}} \qquad (3.5)$$

where a is the species activity and v_i is the species stoichiometric coefficient. Gas activity is equivalent to partial pressure, and for the present calculations, the model assumes H_2O activity equals unity (liquid water). With this assumption, the reversible open circuit voltage for Equation 3.3 can be expressed as

$$E = -\frac{\left(\hat{g}_f\right)_{H_2O} - \left(\hat{g}_f\right)_{H_2} - 0.5\left(\hat{g}_f\right)_{O_2}}{2F} + \frac{RT}{2F} \ln\left(p_{H_2} p_{O_2}^{\frac{1}{2}} \right) \qquad (3.6)$$

This reversible open circuit voltage is the maximum voltage a fuel cell can produce, and the actual voltage will always be less than this maximum due to various loses, called overpotentials, in the system. In the following, three overpotentials, namely activation, ohmic, and mass transport, are discussed in detail.

3.2.1.2 Activation Overpotential

Activation loss is the reduction in cell electrical potential required to increase the current output of a cell beyond its output at electrochemical equilibrium called exchange current density. The equation used to describe the activation losses in the fuel cell model is the Tafel equation (Equation 3.7). The value of 2 in Equation 3.7 is due to the number of charges (electrons) transferred in the hydrogen oxidation reaction (Equation 3.2).

$$\eta_{act} = \frac{RT}{\alpha 2F} \ln\left(\frac{i}{i_o} \right) \qquad (3.7)$$

Equation 3.7 represents a simplification of the Butler–Volmer equation (Equation 3.8).

$$i = i_o \left(e^{\alpha 2 F \eta_{act}/(RT)} - e^{-(1-\alpha)2F\eta_{act}/(RT)} \right) \qquad (3.8)$$

For systems with current densities much larger than the exchange current density (when η_{act} is large), the second term in Equation 3.8 can be neglected, yielding

$$i = i_o e^{\alpha 2F\eta_{act}/(RT)} \tag{3.9}$$

which can be rewritten as Equation 3.7.

The exchange current density, i, represents the current density at electrochemical equilibrium, that is, the point at which the forward and reverse reaction rates are equal.

$$i_1 = i_2 = i_o \tag{3.10}$$

Forward and reverse reaction rates can be written as

$$i_1 = 2FC_R f_1 e^{-\Delta G/(RT)} \tag{3.11}$$

$$i_2 = 2FC_P f_2 e^{-(\Delta G - \Delta G_{rxn} + 2F\Delta\Phi)/(RT)} \tag{3.12}$$

where $C_P i_l$ and $C_R i_l$ are the product and reactant concentrations, respectively, $\Delta G i_l$ is the activation energy for the reaction, and $\Delta\varphi$ is electrical potential at the reaction site. f_1 and f_2 are decay rates and can generally be assumed equal. Setting Equations 3.11 and 3.12 equal to each other and simplifying yields the following relationship:

$$\frac{C_R}{C_P} = e^{(\Delta \hat{g}_f - 2F\Delta\varphi)/(RT)} \tag{3.13}$$

Since $\Delta \hat{g}_f$ $2F$, and R are constants, for a given temperature, we can say that i_o occurs when the electrical potential at the reaction site is proportional to chemical potential:

$$\Delta\varphi \approx \ln\left(\frac{C_R}{C_P}\right) \tag{3.14}$$

Typical values for i_o for PEM fuel cells are on the order of 10^{-4} mA/cm². The operational range of fuel cells is generally many orders of magnitude larger, validating the simplification of the Butler–Volmer equation (Equation 3.9).

The constant α in Equation 3.7, the charge transfer coefficient, describes the asymmetry in activation energy in an electrochemical reaction. Values of αi_l range between 0 and 1, with 0.5 representing a symmetric reaction, that is, when the increase in activation energy for the reverse reaction is equivalent to the decrease in activation energy for the forward reaction. In a fuel cell, α will vary based on type and quantity of catalyst used, and generally has a value between 0.2 and 0.5.

3.2.1.3 Ohmic Overpotential

Moving charged particles incurs losses, and these losses in a fuel cell are referred to as the ohmic overpotential, η_{ohm}. Of the two charged particles transferred in a PEM fuel cell system, protons and electrons, the hydrogen ion transport through the electrolyte accounts for the majority of the resistance. In conducting metals, valence electrons are relatively free to move about the material; in order for ions to move through a material, they must take advantage of free spaces in the physical structure of the electrolyte-vacancies and interstitial sites in ceramics or charged sites in polymers. Ions can also be transported by associating with molecules in a liquid, for example H_3O^+. In any case, charge conductivity is much less for ions than for electrons, and so electrical resistance can be ignored.

The voltage drop due to ion transport resistance then follows Ohm's law:

$$\eta_{ohm} = IR_i \tag{3.15}$$

which can be written in terms of current density as

$$\eta_{ohm} = iAR_i \tag{3.16}$$

3.2.1.4 Mass Transport Overpotential

The physical limits of mass transport rate impose two modes of voltage loss on a fuel cell—decreased Nernst voltage and reaction rate. These will both decrease as reactant concentrations at the reaction site fall away from the bulk flow concentration. The linear concentration gradient that develops can be described by the flux of the reactants, J, as

$$J = -D^{eff} \frac{C_R - C_R^o}{\delta} \tag{3.17}$$

where C_R^o is the reactant concentration in the bulk flow, D^{eff} is the effective diffusivity through the diffusion layer, and δ is the diffusion layer thickness. At steady state, reactant flux through the diffusion layer will equal reactant consumption so that

$$i = 4FJ \tag{3.18}$$

The numerical coefficient in Equation 3.18 is 4 as opposed to 2 because here oxygen is being considered, not hydrogen, and there are four charged particles transferred per O_2 molecule. Since the diffusivity of oxygen is much less than that of hydrogen, the mass transport losses of hydrogen can be neglected.

Combining Equations 3.17 and 3.18 gives a relationship between reaction concentrations and current density (Equation 3.19). From this, the maximum possible current density, when the current density is limited by the diffusivity of the reactant gas and diffusion layer thickness, can be calculated by setting C_R, the reactant concentration at the reactant site, equal to zero (Equation 3.20). This is the limiting current density, or the maximum possible current density of the cell, and is denoted i_l.

$$i = 4FD\frac{C_R^o - C_R}{\delta} \tag{3.19}$$

$$i = 4FD\frac{C_R^o}{\delta} \tag{3.20}$$

The decrease in voltage due to mass transport inefficiencies can be expressed in terms of limiting current density. The concentration (mass transfer) losses resulting from reduced Nernst voltage can be written as the change in Equation 3.5 due to reduced reactant concentration.

$$\eta_{mt,nernst} = \left(E^o + \frac{RT}{nF}\ln C_R^o \right) - \left(E^o + \frac{RT}{nF}\ln C_R \right) = \frac{RT}{nF}\ln\left(\frac{C_R^o}{C_R}\right) \tag{3.21}$$

Equation 3.21 can be combined with Equations 3.19 and 3.20 to find the voltage loss in terms of current density and limiting current density.

$$\eta_{mt,nernst} = -\frac{RT}{nF}\ln\left(1 - \frac{i}{i_l}\right) \tag{3.22}$$

To account for the reduction in reaction rate due to reactant concentration, one must rewrite Equation 3.9 in an alternate format—one that takes into account species concentration.

$$i = i_o\left(\frac{C_R^*}{C_R^o}e^{\alpha 2F\eta_{act}/(RT)}\right) \tag{3.23}$$

Rearranging to get Equation 3.24 in terms of voltage loss,

$$\eta_{act} = \frac{RT}{\alpha nF}\ln\left(\frac{iC_R^o}{i_o C_R^*}\right) \tag{3.24}$$

where C_R^* is an arbitrary reactant concentration. The bulk and actual reactant concentrations can be substituted in for C_R^* and subtracted to find the reduction in voltage.

$$\eta_{mt,act} = \frac{RT}{\alpha nF} \ln\left(\frac{iC_R^o}{i_o C_R} \right) - \frac{RT}{\alpha nF} \ln\left(\frac{iC_R^o}{i_o C_R^o} \right) = \frac{RT}{\alpha nF} \ln\left(\frac{C_R^o}{C_R} \right) \tag{3.25}$$

Equation 3.25 can be rewritten in terms of limiting current density, similar to Equation 3.21.

$$\eta_{mt,act} = -\frac{RT}{\alpha nF} \ln\left(1 - \frac{i}{i_l} \right) \tag{3.26}$$

Equations 3.22 and 3.26 can be combined to give an expression for the total theoretical overpotential resulting from mass transport limitations.

$$\eta_{mt} = -\frac{RT}{nF} \ln\left(1 - \frac{i}{i_l} \right) - \frac{RT}{\alpha nF} \ln\left(1 - \frac{i}{i_l} \right) = -\frac{RT}{nF}\left(1 + \frac{1}{\alpha} \right) \ln\left(1 - \frac{i}{i_l} \right) \tag{3.27}$$

Real fuel cell mass transport losses are often underpredicted by Equation 3.27, and so the equation generally used is

$$\eta_{mt} = -C \ln\left(1 - \frac{i}{i_l} \right) \tag{3.28}$$

Here C is an empirical coefficient that is referred to as the mass transport loss coefficient.

3.2.1.5 Actual Cell Voltage

The actual voltage output of a fuel cell can be modeled by subtracting the overpotentials from the reversible fuel cell voltage.

$$V_{cell} = E - \eta_{act} - \eta_{ohm} - \eta_{mt} \tag{3.29}$$

Substituting Equations 3.6, 3.7, 3.16, and 3.28 into Equation 3.29, the real voltage output of the fuel cell is then written as

$$V_{cell} = \frac{\left(\hat{g}_f \right)_{H_2O} - \left(\hat{g}_f \right)_{H_2} - 0.5\left(\hat{g}_f \right)_{O_2}}{2F} + \frac{RT}{2F} \ln\left(p_{H_2} p_{O_2}^{1/2} \right) - \frac{RT}{\alpha 2F} \ln\left(\frac{i}{i_o} \right)$$

$$- iAR_i + C \ln\left(1 - \frac{i}{i_l} \right) \tag{3.30}$$

TABLE 3.1

Fuel Cell Model Parameters

Symbol	Term	Units
i_o.	Change current density	mA/cm²
	Charge transport coefficient	
R_i	Ionic resistance	Ω
A	Cell active surface area	cm²
i_l	Limiting current density	mA/cm²
	Mass transport loss coefficient	V

Values from Equation 3.30 that are variables in the present model are given in Table 3.1.

3.2.2 Thermal Model

The temperature of the fuel cell stack is determined through a simple heat transfer analysis. The thermal energy generated by the stack is transferred without loss to masses of defined properties. Heat is then dissipated to the environment through convection to a constant temperature fluid.

The amount of thermal energy produced by the fuel cell model is equal to the difference between the ideal and actual power generation of the stack (Zeman, 2010). This is illustrated in Figure 3.2, where Q is the amount of thermal energy transferred to the mass per unit time (Equation 3.31).

$$Q = P^* - P = IE^o - IV_{cell} \qquad (3.31)$$

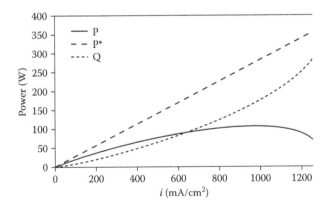

FIGURE 3.2
Fuel cell stack power output (P), ideal power output (P^*), and thermal output (Q).

3.2.3 Humidity Model

3.2.3.1 Modeling Equations

Water content of the fuel cell was determined on a rate basis: water flow rate into the fuel cell with the reactant gasses and water production rate by the cell. Correlations for the effect of humidity on fuel cell properties have been taken from the literature. This section outlines the theoretical basis and modeling equations used to calculate the effect of humidity, changes in temperature, and reactant concentrations on a fuel cell.

3.2.3.1.1 Water Activity

Water activity in the fuel cell was determined from first principles. Since the present fuel cell model was quasi-one-dimensional, humidity distribution was ignored and average water activity was calculated.

Water activity is defined as

$$a \equiv \frac{P_{H_2O}}{P_{H_2O}^{sat}} \tag{3.32}$$

where P_{H_2O} is the actual partial pressure of water vapor in the environment and $P_{H_2O}^{sat}$ is the pressure of water vapor at saturation. For values of $a \leq 1$, water activity is synonymous with relative humidity.

Calculating water activity inside the fuel cell stack begins with a determination of the saturated vapor pressure of the fuel and oxidizer gas streams. This is given in Equation 3.33 (Maggio et al., 2001):

$$P_{H_2O_i}^{sat} = 101,325e^{\left(13.669 - \frac{5096.28}{T_i}\right)} \tag{3.33}$$

where T_i is the average temperature of the inflow gases. The partial pressure of water in the inflow gases is then

$$P_{H_2O_i} = \varphi P_{H_2O_i}^{sat} \tag{3.34}$$

where is the average relative humidity of inflow gases. The total molar flow rate of water into the fuel cell is calculated by assuming that water vapor is an ideal gas, that is,

$$\dot{n}_{H_2O,in} = \frac{P_{H_2O_i}\dot{V}}{RT_i} \tag{3.35}$$

and the total molar flow rate of non-vapor gases into the fuel cell is

$$\dot{n}_{gas,in} = \frac{(P_{avg} - P_{H_2O_i})\dot{V}}{RT_i} \tag{3.36}$$

where P_{avg} is the average of the cathode and anode pressures and $\dot{V}$ is the sum of the cathode and anode gas flow rates.

Besides the water brought into the cell by the fuel and oxidizer streams, there is water produced by the cell's driving electrochemical reaction (Equation 3.3). This is related to the current output of the fuel cell, as each electron produced corresponds to half of a water molecule produced from the hydrogen oxidation reaction. This "half" value is taken into account by the value of 2 in Equation 3.37. Faraday's constant, F, converts the charge per second given by the current into moles per second.

$$\dot{n}_{H_2O,prod} = \frac{I}{2F} \tag{3.37}$$

From Equations 3.35, 3.36, 3.37, and the average of the cathode and anode pressures, the water vapor pressure at the outlet of the cell can be calculated.

$$P_{H_2O} = \left(\frac{\dot{n}_{H_2O,in} + \dot{n}_{H_2O,prod}}{\dot{n}_{gas,in} + \dot{n}_{H_2O,in} + \dot{n}_{H_2O,prod}} \right) P_{avg} \tag{3.38}$$

The average water activity in the cell (Equation 3.39) can be found, using Equation 3.40 for the saturation pressure of water inside the fuel cell.

$$a_{avg} = \frac{P_{H_2O,avg}}{P_{H_2O}^{sat}} = \frac{P_{H_2O_i} + P_{H_2O}}{2 * P_{H_2O}^{sat}} \tag{3.39}$$

$$P_{H_2O_i}^{sat} = 101,325e^{\left(13.669 - \frac{5096.28}{T}\right)} \tag{3.40}$$

Here T is the fuel cell temperature in Kelvin.

3.2.3.1.2 Membrane Resistance

The model used to take into account the effects of water activity on membrane resistance was based on the work by Springer et al. (1991). Figure 3.3 shows the relationship between water vapor activity in the environment and water content of the membrane at 303 K. Water content of the membrane is defined as the number of molecules of water per sulfonic acid $(SO_3^- H^+)$ site in the membrane.

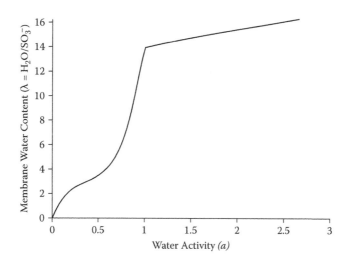

FIGURE 3.3
Relationship between water vapor activity and membrane water content at 303 K. (Plot generated using a model from Springer, T.E., Zawodzinski, T.A., and Gottesfeld, S., *Journal of the Electrochemical Society*, 138: 2334–2342, 1991.)

The relationship between a_{avg} and λ_{303} is a piecewise function given by Equations 3.41 and 3.42.

$$\lambda_{303} = 0.043 + 17.81a_{avg} - 39.85a_{avg}^2 + 36.0a_{avg}^3 \quad \text{for} \quad 0 < a_{avg} \le 1 \quad (3.41)$$

$$\lambda_{303} = 14 + 1.4(a_{avg} - 1) \quad \text{for} \quad 1 \le a_{avg} \le 3 \quad (3.42)$$

For $\lambda > 1$, water content was related to membrane conductivity (Springer et al., 1991) as

$$\sigma_{303} = 0.005193\lambda_{303} - 0.00326 \quad \text{for} \quad \lambda > 1 \quad (3.43)$$

and correlated to temperature change by

$$\sigma = \sigma_{303}e^{\left[1268\left(\frac{1}{303} - \frac{1}{T}\right)\right]} \quad (3.44)$$

Membrane resistance is related to conductivity by the following definition:

$$\Omega \equiv \frac{d}{A\sigma} \quad (3.45)$$

3.2.3.1.3 Limiting Current Density

The limiting current density of the fuel cell, i_l, is also affected by the relative humidity of the inflow gasses. As water content of the cathode gases increases, the bulk concentration of oxygen is decreased. For the reasons given in the discussion of mass transport overpotential, only oxygen is considered here, and the effect of hydrogen concentration on the limiting current density is assumed to be negligible. Limiting current density is related to bulk concentration by Equation 3.46.

$$i_l = 4FD^{eff}\left(\frac{\overline{C}_R^0}{\delta}\right) \tag{3.46}$$

D^{eff} is the effective diffusivity of the oxygen in the gas diffusion layer, and it is proportional to the bulk diffusivity, D_{AB}, but modified by the geometry of the diffusion layer. The temperature and pressure dependence of D_{AB} has been given by Fuller et al. (1966) as

$$D_{AB} = \frac{10^{-3}T^{7/4}\left(\frac{1}{M_A}+\frac{1}{M_B}\right)^2}{P\left[\left(\sum_A v_i\right)^{1/3}+\left(\sum_B v_i\right)^{1/3}\right]^2} \tag{3.47}$$

From this, the percent change in diffusivity due to a temperature and pressure change from reference values is given by

$$\frac{D_{AB(T,P)} - D_{AB(T_{ref},P_{ref})}}{D_{AB(T_{ref},P_{ref})}} = \left(\frac{T}{T_{ref}}\right)^{7/4}\left(\frac{P_{ref}}{P}\right) - 1 \tag{3.48}$$

Consequently, given a reference effective diffusivity at a known temperature and pressure, the effective diffusivity at any temperature and pressure can be found.

$$D^{eff} = \left(\frac{T}{T_{ref}}\right)^{7/4}\left(\frac{P_{ref}}{P}\right)D^{eff}_{(T_{ref},P_{ref})} \tag{3.49}$$

δ in Equation 3.46 is the diffusion layer thickness, and the value of 4 represents the number of charged particles transferred in the cell per O_2 molecule. The bulk concentration, $\overline{C}_R^0$, can be derived from a ratio of the molar and volumetric flow rates of oxygen.

$$\overline{C}_R^0 = \frac{\dot{n}_{O_2}}{\dot{V}} = X_{O_2 i}\frac{\dot{n}_{gas}}{\dot{V}} \tag{3.50}$$

where X_{O_2i} is the mole fraction of dry inflow gas, and

$$\frac{\dot{n}_{gas}}{\dot{V}} = \frac{(P_{catch} - P_{H_2O,avg})}{RT} \tag{3.51}$$

The accumulation of liquid water in the gas diffusion layer and the subsequent reduction in effective diffusivity was not considered here. It is assumed that, for relative humidity levels under 100%, gas diffusion layer saturation is negligible (Dawes et al., 2009). Operation of a fuel cell near, at, or above 100% relative humidity for any length of time is not recommended, as liquid water blocks catalyst sites and reduces the effective porosity of the gas diffusion layer (Yamada et al., 2006).

3.2.3.1.4 Reactant Mole Fraction

The reactant concentration in the humidified anode and cathode streams must also be calculated for determination of the Nernst voltage (Equation 3.6). The values of the average oxygen and hydrogen mole fractions are given in Equations 3.52 and 3.53, respectively.

$$X_{O_2} = X_{O_2i} \left(1 - \frac{P_{H_2O,avg}}{P_{cath}} \right) \tag{3.52}$$

$$X_{H_2} = X_{H_2i} \left(1 - \frac{P_{H_2O,avg}}{P_{anode}} \right) \tag{3.53}$$

This effect on the Nernst voltage is distinct from the effect given by Equation 3.22; Equations 3.52 and 3.53 account for a change in open circuit voltage caused by a reduction in bulk concentration, whereas Equation 3.22 accounts for a reduction in voltage due to reduced local reactant concentration at high current densities.

3.2.3.1.5 Exchange Current Density

The change in exchange current density, i_o, with temperature was noted by Rajani and Kolar (2007), who described the relationship as

$$i_{o(T)} = i_{o(T_{ref})} e^{\left[\frac{\Delta E}{R} \left(\frac{1}{T_{ref}} - \frac{1}{T} \right) \right]} \tag{3.54}$$

where ΔE is the reaction activation energy and has a value of 72,000 (J/mol) and T_{ref} is the reference temperature from Equation 3.48.

3.3 Model Implementation

Incorporation of humidity dynamics into the fuel cell model was achieved through a calculation loop external to the fuel cell model. Figure 3.4 shows the complete setup. Values output by the fuel cell model, namely temperature and current, were accepted as inputs and the humidity equations output membrane resistance, limiting current density, and reactant mole fractions for use by the fuel cell model. The values were returned to the fuel cell model by first recording them as a Results Variable (RLT), then by reading those values with an RLTDependence—a dependent variable that takes its value from its an RLT. The parameters in the fuel cell model were then set to read their associated RLTDependence. Since the RLTDependences require an initial output value, the fuel cell power request was delayed by 10 times the RLT sample rate. This allows a true calculated output value to be determined before fuel cell operations begin.

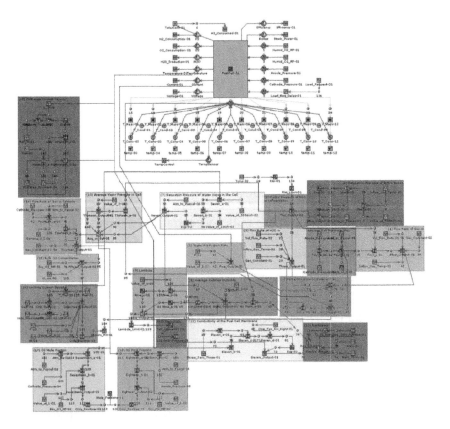

FIGURE 3.4
Fuel cell humidity model implemented in GT-Suite®.

TABLE 3.2

New User Input Parameters for the Humidity Model

Symbol	Description	Units
P_{cath}	Pressure of cathode chamber	Pa
P_{anode}	Pressure of anode chamber	Pa
Φ	Relative humidity of the inlet stream	—
$\dot{V}$	Total (sum) volumetric flow rate of the fuel and oxidizer streams	m³/s
T_i	Temperature in the inlet stream	K
T_{ref}	Reference temperature at which the FC was modeled	K
X_{O2i}	Mole fraction of oxygen in the dry stream	—
X_{H2i}	Mole fraction of hydrogen in the dry stream	—
D	Membrane thickness	cm
A	Fuel cell active surface area	cm²
D_{ref}^{eff}	Effective diffusivity of oxygen through the gas diffusion layer	cm²/s
δ	Thickness of the gas diffusion layer	cm

New parameters introduced by the humidity modeling equations that are required to be specified by the user are listed in Table 3.2.

3.4 Results and Discussion

This section begins by discussing the significance of the polarization curve and what effect different modeling parameters and their associated physical phenomena have on that curve. The ability of the standard model to reproduce a fuel cell and its predictive capabilities are then explored. Finally, the fuel cell model with the additional relationships discussed before is compared with experimental data.

3.4.1 Parametric Variations

It cannot be assumed that all parameters of the fuel cell model will be known; C, the mass transport loss coefficient, is a purely empirical value and other parameters, such as the limiting current density, i_l, are dependent on many disparate factors (gas diffusion layer mesh type, initial pore size, and percent compression for i_l) and, thus, cannot easily be determined theoretically and may be difficult to measure in situ. Because of this, the values of the parameters are found by reproducing the polarization curve of the fuel cell to be modeled.

The polarization curve of a fuel cell is a plot of how the voltage varies over its range of current outputs (Figure 3.5). The three overpotentials (Equations

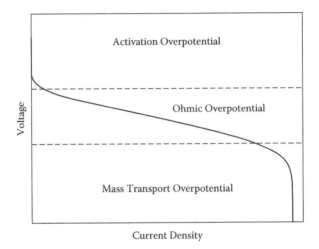

FIGURE 3.5
A generic polarization curve showing the overpotential regions.

3.7, 3.16, and 3.28) dominate different regions of the polarization curve, as can be seen in Figure 3.6—the activation overpotential predominates in the low current density region, mass transport losses have their greatest impact at high current densities, and ohmic losses increase proportionally with increasing current density. The effect of the overpotentials can be deconstructed further with an analysis of the effects of changes in individual parameters. Figures 3.7 through 3.11 show the change in fuel cell performance resulting from variation in a single parameter. With an understanding of this, it is

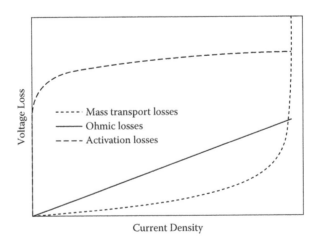

FIGURE 3.6
Voltage losses contributing to a polarization curve.

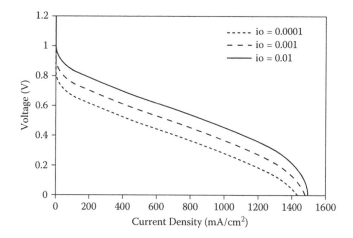

FIGURE 3.7
Variable exchange current density (mA/cm²).

possible to qualitatively replicate a polarization curve and therefore model the fuel cell that produces the polarization curve.

3.4.2 Validation of Non-humid PEM Fuel Cell Model

The first direct comparison of the non-humid fuel cell model under consideration here was performed against data presented by Fontes et al. (2007). The Fontes model was presented alongside values derived from the fuel cell

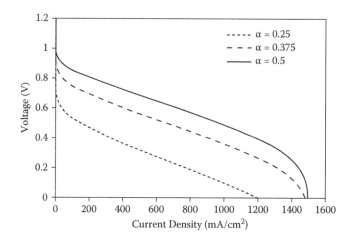

FIGURE 3.8
Variable charge transfer coefficient.

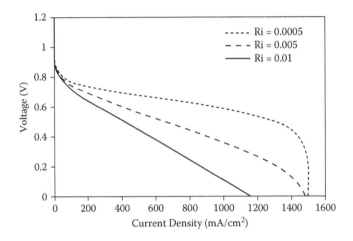

FIGURE 3.9
Variable membrane resistance (Ω).

against which they tested their model. From these values, it was possible to derive the parameters used by the model in the present study to define a fuel cell's performance. This allowed it to be seen if the model was capable of reproducing data without using the parameters as "fitting coefficients," chosen specifically to reproduce a polarization curve (Pisani et al., 2002b). A comparison between the two data sets is given in Figure 3.12. Values for the modeling parameters are listed in Table 3.3.

These results give confidence that the model output is a reliable interpretation of its input parameters and that, for a set of parameters, it is consistent

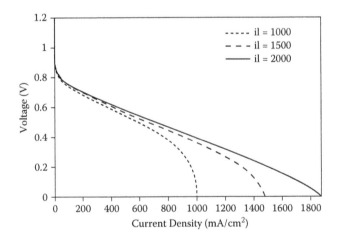

FIGURE 3.10
Variable limiting current density (mA/cm²).

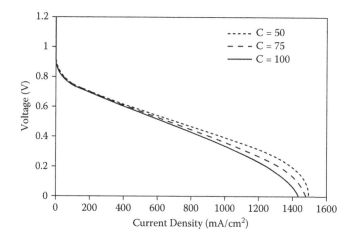

FIGURE 3.11
Variable mass transport loss coefficient (mV).

with fuel cell performance in the literature. This also implies that, inversely, when a polarization curve is copied with parameters used as fitting coefficients, these parameters are meaningful with respect to the electrochemical theory and not arbitrary.

This was confirmed with a comparison to the model results given in O'Hayre et al. (2007) (Figure 3.13). In that study, mass transport loses were considered differently than they have been here, and so values for i_l and C were not derivable from reported modeling parameters. In lieu of this, these parameters were used as fitting coefficients; values were selected such that

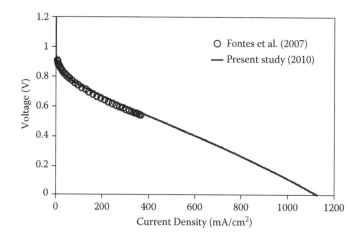

FIGURE 3.12
Comparison of fuel cell outputs given by Fontes et al. (2007) and the present study's model.

TABLE 3.3

Modeling Parameters Given by Fontes et al. (2007)

Parameter	Value	Units
i_o	0.00234	mA/cm^2
i_l	2000	mA/cm^2
R_i	0.0041	Ω
α	0.35	—
C	0.562	V

Source: Data from Fontes, G., Turpin, C., Astier, S., and Meynard, T.A., IEEE Transactions on Power Electronics, 22: 670–678, 2007.

the performance of the model discussed here matched that of the O'Hayre model. This served to confirm that the mass transport modeling equation (Equation 3.28) gives a polarization curve contour consistent with that of models reported in the literature.

The next comparison made was with a study by Ersoz et al. (2006). In that study, a fuel cell system was modeled and the efficiency for a variety of fuel cell stack sizes was reported. Since only the number of cells in the fuel cell stack changed, the voltage/current response (the polarization curve) for each cell did not change, but the current and voltage output needed by each stack size to meet the requested power output (constant for all stack sizes) did change, as illustrated by Figure 3.14. This allowed this study's calculations of efficiency for various points on the polarization curve to be compared. The

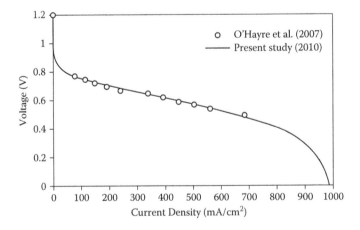

FIGURE 3.13
Polarization curves from O'Hayre et al. (2007) and the present study.

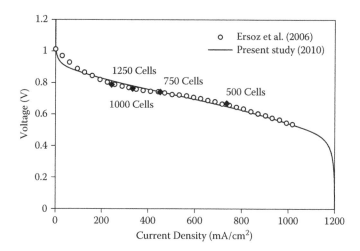

FIGURE 3.14
Polarization curve comparison with Ersoz et al. (2006), showing current/voltage output for given stack sizes.

results of this can be seen in Table 3.4, and they show good correlation with the values derived by Ersoz et al. (2006). This shows that the model's calculation of efficiency, and hence of the rate of energy transferred to the thermal model, is accurate.

Further comparisons were made to test the model's ability to account for a change in reactant pressure. Similar to the Ersoz et al. comparison, the model parameters were adjusted so that the model output fit fuel cell data presented by Kim et al. (1995) at 1 atm pressure. The model pressure was then changed while all other inputs and parameters were held constant. The resulting output as compared to actual fuel cell data presented by Kim et al. can be seen in Figure 3.15. Clearly, the model somewhat failed to accurately accommodate the change in pressure. The pressure change is only accounted for by the Nernst Equation (Equation 3.6). This ignores the change in bulk reactant concentration and the effect that has on the mass transport losses in the cell (Equations 3.46, 3.50, and 3.51).

Kim et al. also presented data for the same fuel cell at an increased temperature. The comparison between the fuel cell model and the experimental data

TABLE 3.4

Comparison of Calculated Efficiencies of Various Fuel Cell Stack Sizes

Number of Cells	1250	1000	750	500
Ersoz et al. (2006)	64.6	62.6	60.5	54.2
Present Study	66.5	64.6	61.5	54.5
% Difference	2.9	3.2	1.7	0.48

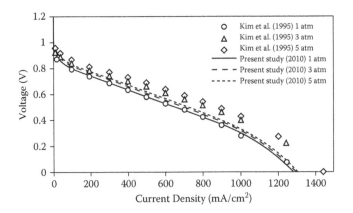

FIGURE 3.15
Comparison of polarization curves at various reactant pressures.

can be seen in Figure 3.16. The model shows a decreased voltage response at higher temperatures; this is due to the increase in activation losses as temperatures increase (Equation 3.7). The experimental data, however, shows a more complex interaction. From observation of Figure 3.16 in comparison with Figures 3.9–3.11, it appears that the mass transport and ohmic overpotentials decrease with increasing temperature in a real fuel cell. It should be noted that the fuel cell model parameters can be set as temperature-dependent arrays, but this was not considered since the main attraction of this model is its ease and speed of use, and creating arrays would require fitting multiple polarization curves over a range of temperatures.

Kim et al. presented data on reduced oxygen concentrations, and this can be used to corroborate the findings from Figure 3.16. Figure 3.17 shows that,

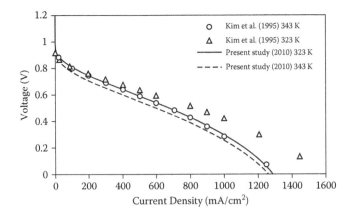

FIGURE 3.16
Comparison against fuel cell data at multiple temperatures at 1 atm.

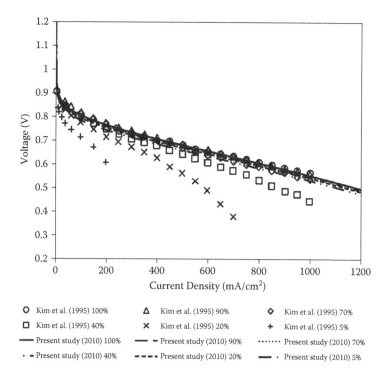

FIGURE 3.17
Fuel cell response at variable oxygen concentrations. Percent given is $\%O_2$ in the oxygen/argon cathode feed.

as with the change in pressure, the model accounts for the change in Nernst voltage but does not take into account the increased mass transport losses that accompany a decrease in bulk reactant concentration.

3.4.3 Validation of Humid PEM Fuel Cell Model

The PEM fuel cell model with the additional relationships detailed in Section 3.2.3 shows substantial improvement over the standard model, though some of the results are still not ideal.

Figure 3.18 shows the same comparison as Figure 3.15, but with the new model. The model predictions are more accurate, especially at high current densities, though it appears that the model still under-predicts the decrease in mass transport losses at higher pressures. The inverse pressure relationship of D^{eff} in Equation 3.49 is a relationship that has been confirmed time and again, and if the ideal gas assumption made in Equation 3.51 is valid, this leaves the size of the diffusion layer as only possible source of the error. The majority of the diffusion layer is composed of the gas diffusion layer (GDL), typically a carbon cloth that forms an electrical connection between the catalyst and

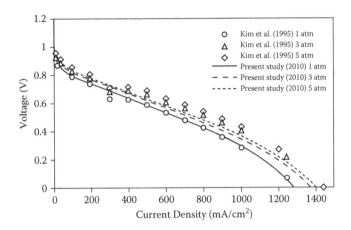

FIGURE 3.18
Comparison of humid model predictions with experimental data for the effects of pressure change.

current collector, but the diffusion layer extends slightly beyond the GDL into the boundary layer of the gas stream. At increased pressures the viscosity of the gas stream increases, which in turn decreases the Reynolds number. At lower values of Reynolds number, the slope of the velocity in the boundary layer decreases, which signifies a reduction in the percent of the flow channel at which diffusion is significant. This decrease in diffusion layer size and the corresponding decrease in mass transport losses is not accounted for in the model and could be responsible for the discrepancy in Figure 3.17.

Figure 3.19 shows the same comparison as Figure 3.16, substituting the new model for the standard model. While the new comparison shows greater accuracy in the low current density region, mainly due to the increased exchange current density at elevated temperatures (Equation 3.54) offsetting the decrease in voltage observed in Figure 3.16, the results do not show great parity at medium to high current densities and there are two possible reasons for this.

The model may be under-predicting the decrease in membrane resistance with respect to a temperature increase. The membrane used by Kim et al. was Nafion 115, which, while similar to the Nafion 117 membrane studied by Springer et al. (resulting in Equations 3.41–3.44), has been shown by Yang et al. (2004) to follow a water uptake trend that differs from that of Nafion 117. This implies that the resistance response of the membrane to changes in temperature is also different, but this question has not been addressed in the fuel cell literature.

Probably more significant is the disparity in apparent mass transport losses between the experimental and modeled data in Figure 3.19. The model predicts an overall reduction in limiting current density and hence an increase in mass transport losses; the reduction in bulk concentration due to increased temperature and water vapor pressure (required to maintain humidity levels

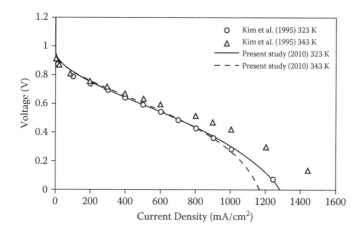

FIGURE 3.19
Comparison of humid model predictions with experimental data for the effects of temperature change.

at the increased temperature) in Equation 3.51 overwhelms the increase in effective diffusivity at increased temperatures implied by Equation 3.49. This is not seen in the experimental data, implying that there is a phenomenon at work that has not been taken into account by the model. Zhou et al. (2009) reported on membrane swelling increasing due to increased temperature and relative humidity and suggested that "GDL deformation reduces gas flow area and through-plane thickness, which can facilitate the gas flow and therefore reduce flow resistance." They also stated that membrane swelling reduces contact resistance between the GDL and current collector, reducing ohmic losses.

Error could also be attributable to a discrepancy between reported fuel cell temperature and average cell temperature. The fuel cell model presented here considers the average membrane temperature of the fuel cell, but Le et al. (2008) illustrated clearly that there exists a great variation in temperature across the membrane. It was assumed that the temperature reported was an average, but this may not be correct.

The new model predictions seen in Figure 3.20 are much improved from those of the standard model predictions from Figure 3.17. The model shows the same trend of increased mass transport losses at low reactant concentrations that the experimental data shows; however, this increase in losses is overpredicted by the model. The model does not take into account the change in diffusivity of oxygen as the percent of argon increases, but if this were taken into account, it would result in lower oxygen diffusivities (Fuller et al., 1966) and consequently less accurate model predictions. Argon, being a noble gas, is most likely not causing a change in the GDL thickness, and the difference in viscosity between it and oxygen is unlikely to be responsible for so large a deviation between predicted and actual mass transport losses. If the ideal gas assumption in Equation 3.51 is valid, then there is nothing in the theory laid out

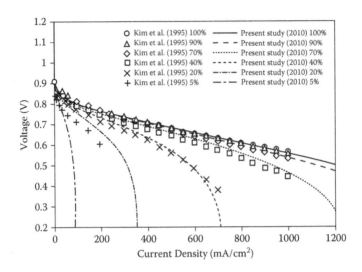

FIGURE 3.20
Comparison of humid model predictions with experimental data for the effects of changing oxygen concentration in an O_2/Ar mixture.

in Section 2.3.1.3 for predicting mass transport losses that should be suspected of error. Consequently, the source of the discrepancy between the experimental and predicted polarization curves in Figure 3.20 remains an open question.

Figure 3.21 compares model predictions for the effects of a change in the relative humidity of inflow gasses. Here the polarization curve was matched to data collected at 70% relative humidity by Yan et al. (2006). When the humidity is increased to 100%, the mass transport losses increase due to oxygen being displaced by water vapor in the cathode, but this is offset at lower current densities by the decreased membrane resistance that results from increased water content of the membrane. The increased diffusivity of oxygen through water vapor is not accounted for by the model, but this would increase the predicted limiting current density by less than 3%—not enough to account for the discrepancy between the modeled and experimental data. Another possible source of the disagreement between the two data sets lies with the membrane resistance and how polarization curves are generated experimentally. With a physical system, the current is increased from open circuit to short circuit. As the current increases, the amount of water produced by the cell increases in accordance with Equation 3.37. Liu and Wu (2006) showed that this reduced the membrane resistance as the test was run; instead of the straight line for ohmic overpotential shown in Figure 3.6, the slope would decrease as the current density increased. The modeling software, however, generates a polarization curve as a snapshot, using a single value for resistance. The result of this is that a polarization curve from a physical fuel cell would exhibit a higher voltage output than a model would

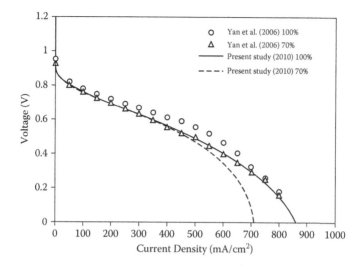

FIGURE 3.21

Comparison between predicted and reported performance after a change in relative humidity from 70% to 100%.

as current density increased, but at high current densities this effect would be overwhelmed by the mass transport losses. This could partially explain the profile of the 100% relative humidity curve presented by Yan et al. (2006) and the discrepancy between it and the model prediction.

Yan et al. (2006) also reported data that allowed the use of average relative humidity in Equation 3.34 to be tested. Figure 3.22 compares the model prediction for an average relative humidity of 85% with two data sets: one with the cathode stream humidified to 70% and the anode to 100%, and another with cathode at 100% and the anode at 70%. The average relative humidity of both of these data sets is 85%, and, if the average relative humidity assumption is accurate, the two should match, but, as can be seen, they are not equal and the model prediction falls between them.

3.5 Summary and Conclusions

After discussing the operational principles of PEM fuel cells, a simple model was developed and incorporated into a commercial software package that has been widely used for predicting the performance of conventional gasoline and diesel engines. The model's implementation in an integrated computational package that is already available can allow all relevant subsystems to be simulated in the same environment, which could be attractive to industries wishing to incorporate fuel cells into practical various applications.

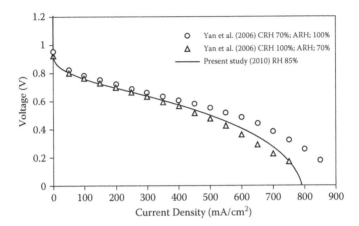

FIGURE 3.22
Comparison between non-symmetric humidification and the average humidity approximation.

This is especially important for hybrid vehicles where fuel cells can power electric motors or be combined with advanced batteries.

The simple fuel cell model was discussed and evaluated against independent data reported in the literature for its suitability to predict the performance of PEM fuel cells. It was found to be capable of reproducing the voltage/current response of actual fuel cells as well as the change in efficiency resulting from a change in current output from those cells. The model was shown to have insufficient predictive abilities when the conditions were changed. The model also had no mechanism to account for the effects of relative humidity. A model to relate the limiting current density to a change in bulk reactant concentration due to changes in water vapor content, temperature, pressure, and inlet oxygen mole fraction was therefore implemented. These remedied many of the deficiencies of the standard model and allowed for it to respond more accurately to variable fuel cell operating conditions. The findings reported here established this model as capable of simulating PEM fuel cells with a reasonable degree of accuracy and the low computational intensity inherent to analytical modeling. Given the software environment the model is implemented in, this could be of significant aid to the design and optimization of fuel cell- and hybrid-powered vehicles.

References

Barley, C.D., and Gawlik, K. (2009). Analysis of buoyancy driven ventilation of hydrogen from buildings. *International Journal of Hydrogen Energy*, 34: 5592–5603.

Bernardi, D.M., and Verbrugge, M.W. (1992). A mathematical model of the solid-polymer-electrolyte fuel cell. *Journal of the Electrochemical Society*, 138: 2477–2491.

Berning, T., Lu, D.M., and Djilali, N. (2002). Three-dimensional computational analysis of transport phenomena in a PEM fuel cell. *Journal of Power Sources*, 106: 284–294.

Biyikoglu, A. (2005). Review of proton exchange membrane fuel cell models. *International Journal of Hydrogen Energy*, 30: 1181–1212.

Buchi, F.N., and Srinivasan, S. (1997). Operating proton exchange membrane fuel cells without external humidification of the reactant gases. *Journal of the Electrochemical Society*, 144: 2767–2772.

Crowl, D.A., and Jo, Y. (2007). The hazards and risks of hydrogen. *Journal of Loss Prevention in the Process Industries*, 20(2): 158–164.

Dahoe, A.E., and Molkov, V.V. (2006). On the development of an international curriculum on hydrogen safety engineering and its implementation into educational programmes. *International Journal of Hydrogen Energy*, 32: 1113–1120.

Dawes, L.E., Haspal, N.S., Family, N.S., and Turan, A. (2009). Three-dimensional CFD modeling of PEM fuel cells: an investigation into the effects of water flooding. *Chemical Engineering Science*, 64: 2781–2794.

Dutta, S., Shimpalee, S., and Van Zee, L.W. (2000). Three-dimensional numerical simulation of straight channel PEM fuel cells. *Journal of Applied Electrochemistry*, 30: 135–146.

Ersoz, A., Olgun, H., and Ozdogan, S. (2006). Simulation study of a proton exchange membrane (PEM) fuel cell system with autothermal reforming. *Energy*, 31: 1490–1500.

Fontes, G., Turpin, C., Astier, S., and Meynard, T.A. (2007). Interactions between fuel cells and power converters: influence of current harmonics on a fuel cell stack. *IEEE Transactions on Power Electronics*, 22: 670–678.

Fuller, E.N., Schettler, P.D., and Giddings, J.C. (1966). A new method for prediction of binary gas-phase diffusion coefficients. *Industrial and Engineering Chemistry*, 58: 18–27.

Fuller, T.F., and Newman, J. (1993). Water and thermal management in solid-polymer electrolyte fuel cells. *Journal of the Electrochemical Society*, 140: 1218–1225.

Goswami, D., Mirabal, S., Goel, N., and Ingley, A. (2003). A review of hydrogen production technologies. Fuel Cell Science, Engineering and Technology: First International Conference on Fuel Cell Science, Engineering and Technology, USA, 61–74.

Gupta, S., Brinster, J., Studer, E., and Tkatschenko, I. (2009). Hydrogen related risks within a private garage: concentration measurements in a realistic full scale experimental facility. *International Journal of Hydrogen Energy*, 34: 5902–5911.

Kim, J., Lee, S.M., and Srinivasan, S. (1995). Modeling of proton exchange membrane fuel cell performance with an empirical equation. *Journal of the Electrochemical Society*, 142: 2670–2674.

Koylu, U.O., Vudumu, S.K. and Sheffield, J.W. (2009). Hydrogen Safety in Accidental Release Scenarios. Missouri Energy Summit, Columbia, MO, April 2009.

Le, A.D., and Zhou, B. (2008). A general model of proton exchange membrane fuel cell. *Journal of Power Sources*, 182: 197–222.

Liu, Q., and Wu, J. (2006). Multi-resolution PEM fuel cell model validation and accuracy analysis. *Journal of Fuel Cell Science and Technology*, 3: 51–61.

MacIntyre, I., Tchouvelev, A.V., Hay, D.R., Wong, J., Grant, J., and Benard, P. (2007). Canadian hydrogen safety program. *International Journal of Hydrogen Energy*, 32: 2134–2143.

Maggio, G., Recupero, V., and Pino, L. (2001). Modeling polymer electrolyte fuel cells: an innovative approach. *Journal of Power Sources*, 101: 275–286.

Mazumder, S., and Cole, J.V. (2003). Rigorous 3-D mathematical modeling of PEM fuel cells. *Journal of the Electrochemical Society*, 150: A1503–A1509.

Meng, H., and Wang, C.Y. (2004a). Electron transport in PEMFCs. *Journal of the Electrochemical Society*, 151: A358–A367.

Meng, H., and Wang, C.Y. (2004b). Large-scale simulation of polymer electrolyte fuel cells by parallel computing. *Chemical Engineering Science*, 59: 3331–3343.

Murgia, G., Pisani, L., Valentini, M., and D'Aguanno, B. (2002). Electrochemistry and mass transport in polymer electrolyte membrane fuel cells. *Journal of the Electrochemical Society*, 149: A31–A38.

National Aeronautics and Space Administration: Office of Safety and Mission Assurance. Safety Standard for Hydrogen and Hydrogen Systems. NSS 1740.16.

Nguyen, T.V., and White, R.E. (1993). A water and heat management model for proton-exchange-membrane fuel cells. *Journal of the Electrochemical Society*, 140: 2178–2186.

O'Hayre, R., Fabian, T., Lister, S., Prinz, F.B., and Santiago, J.G. (2007). Engineering model of a passive planar air breathing fuel cell cathode. *Journal of Power Sources*, 167: 118–129.

Pisani, L., Murgia, G., Valentini, M., and D'Aguanno, B. (2002a). A working model of polymer electrolyte fuel cells. *Journal of the Electrochemical Society*, 149: A898–A904.

Pisani, L., Murgia, G., Valentini, M., and D'Aguanno, B. (2002b). A new semi-empirical approach to performance curves of polymer electrolyte fuel cells. *Journal of Power Sources*, 108: 192–203.

Rajani, B.P.M. and Kolar, A.K. (2007). A model for a vertical planar air breathing PEM fuel cell. *Journal of Power Sources*, 164: 210–221.

Russell, P. (2009, October 25). Passengers aboard LZ 129 Hindenburg—May 3–6, 1937. Message posted to http://facesofthehindenburg.blogspot.com/2009/10/passengers-aboard-lz-129-hindenburg-may.html

Sandia National Laboratories 2007. Hydrogen safety, codes, and standards. Retrieved from http://www.ca.sandia.gov/8700/projects/content.php?cid=183

Sharifi Asl, S.M., Rowshanzamir, S., and Eikani, M.H. (2010). Modelling and simulation of the steady-state and dynamic behavior of a PEM fuel cell. *Energy*, 35: 1663–1646.

Springer, T.E., Zawodzinski, T.A., and Gottesfeld, S. (1991). Polymer electrolyte fuel cell model. *Journal of the Electrochemical Society*, 138: 2334–2342.

Vudumu, S.K., and Koylu, U.O. (2009a). A computational study on performance, combustion and emission characteristics of a hydrogen-fueled internal combustion engine. *Proceedings of ASME IMECE2009* -11183.

Vudumu, S.K., and Koylu, U.O. (2009b). Detailed simulations of the transient hydrogen mixing, leakage and flammability in air in simple geometries. *International Journal of Hydrogen Energy*, 34: 2824–2833.

Wahiduzzaman, S., Kolade, B., and Buyuktur, S. (2004). An integrated proton exchange membrane fuel cell vehicle model. *SAE Technical Paper Series*, 2004-01-1474.

Wang, Z.H., Wang, C.Y., and Chen, K.S. (2001). Two-phase flow and transport in the air cathode of proton exchange membrane fuel cells. *Journal of Power Sources*, 94: 40–50.

Yamada, H., Hatanaka, T., Murata, H., and Morimoto, Y. (2006). Measurement of flooding in gas diffusion layers of polymer electrolyte fuel cells with conventional flow field. *Journal of the Electrochemical Society*, 153: A1748–A1754.

Yan, Q., Toghiani, H., and Causey, H. (2006). Steady state and dynamic performance of proton exchange membrane fuel cells (PEMFCs) under various operating conditions and load changes. *Journal of Power Sources*, 161: 492–502.

Yang, C., Srinivasan, S., Bocarsly, A.B., Tulyani, S., and Benziger, J.B. (2004). A comparison of physical properties and fuel cell performance of Nafion and zirconium phosphate/Nafion composite membranes. *Journal of Membrane Science*, 237: 145–161.

Zawodzinski, T.A., Neeman, M., Sillerud, L.O., and Gottesfeld, S. (1991). Determination of water diffusion coefficients in perfluorosulfonate ionomeric membranes. *Journal of Physical Chemistry*, 95: 6040–6044.

Zeman, J. Personal correspondence, Febuary 17, 2010.

Zhou, Y., Lin, G., Shih, A.J., and Hu, S.J. (2009). Assembly pressure and membrane swelling in PEM fuel cells. *Journal of Power Sources*, 192: 544–551.

4

Dynamic Modeling and Control of PEM Fuel Cell Systems

Lie Tang, Nima Lotfi, Joseph Ishaku, and Robert G. Landers

CONTENTS

4.1 Introduction

This chapter discusses the basic principles of fuel cells including the history and different types of fuel cells along with their properties, structure, and applications, with a special focus on Polymer Electrolyte Membrane (PEM) fuel cells. Auxiliary devices needed for safe and efficient operation of PEM fuel cells are also introduced. Some well-known control-oriented dynamic models of PEM fuel cell components are presented. Simulation analysis of a typical PEM fuel cell is conducted based on the dynamic control-oriented models. Finally, commonly used control algorithms, such as oxygen excess ratio and temperature regulation, are presented and implemented using the control-oriented models.

4.1.1 Fuel Cell Fundamentals

Fuel cells are devices that employ an electrochemical process to convert the chemical energy of a fuel into electrical energy. Output electrical energy can be used as a power source in vehicles, electronic devices, household applications, and so on, or as a backup power source for electrical grids. There has been an increasing interest in the last few decades in fuel cell technologies due to their high efficiency and environmental friendliness. In comparison to an energy storage device like a battery, a fuel cell converts chemical energy of a fuel into electrical energy without requiring storage materials in its structure. Fuel cells are also different from conventional heat engines in that they generate electricity directly from chemical energy, whereas conventional heat engines have an intermediate stage where chemical energy is converted into mechanical energy.

The basic fuel cell operating principles were first discovered in 1839 by British physicist Sir William Grove. However, the first practical fuel cells were developed and implemented in 1950 by Francis Bacon, a British scientist. Later, more practical applications of fuel cells were demonstrated in U.S. space programs (Nehrir et al., 2009). Fuel cells have become more viable for different applications and are increasingly being developed and commercialized. However, some fundamental obstacles must be overcome to achieve the full potential of fuel cells. These obstacles include fuel availability, reducing their cost, and improving their operating reliability.

As examples of practical fuel cell applications, the first fuel cell powered buses were introduced in 1993 by Ballard Power Systems (Gou et al., 2010). In the late 1990s and the early 2000s, major automotive companies developed fuel cell vehicle prototypes. These prototypes have been subjected to numerous tests in different locations such as the United States, Japan, and Europe. In addition to automotive applications of fuel cells, more than 2500 fuel cells have been employed as stationary power sources in offices, hospitals, houses, utility power plants, and so on. Fuel cells have also been used in portable power applications originating from prototypes introduced by Samsung Electronics in 2005 (Gou et al., 2010).

The recent increased interest in commercializing fuel cells is due to their numerous advantages. These advantages include clean by-products (e.g., water), meaning they are nearly zero emission energy devices, with only low emissions of nitrogen and sulfur oxides. They are also quiet energy production devices due to the lack of moving parts. Furthermore, fuel cells produce higher power density and efficiency, around 40% electric efficiency, than traditional engine/generator combinations. Also, in large-scale fuel cells, the "waste" heat can be used for heating purposes, which, in turn, will increase the overall system efficiency. Finally, as different types of fuel cells use various conventional and alternative fuels such as hydrogen, ethanol, methanol, and natural gas, which can be generated from renewable energy sources, the dependence on oil for mechanical and electrical energy production will decrease.

4.1.2 Types of Fuel Cells

Fuel cells are generally categorized according to their electrolyte type, which determines their operating temperature and the type of fuel required (Nehrir et al., 2009). Although the chemical reactions taking place inside different types of fuel cells might be slightly different, they all originate from the same electrochemical reaction. Due to their different characteristics, each type of fuel cell is suitable for specific applications.

Fuel cells that are considered low-temperature usually perform below 200°C. At higher temperatures, vapors that might be produced can damage the fuel cell electrolyte. The most common types of low-temperature fuel cells are alkaline fuel cells (AFC), phosphoric acid fuel cells (PAFC), and polymer electrolyte membrane fuel cells (PEMFC). The catalyst used in these types of fuel cells (usually platinum) is very sensitive to carbon monoxide (CO), so pure hydrogen must be used as their fuel. Additional fuel reforming systems are required if pure hydrogen is not available. This is a disadvantage of the low-temperature fuel cells.

High-temperature fuel cells have the capability of converting poisonous contaminants such as CO to hydrogen. They can even oxidize hydrocarbon fuels (e.g., CH_4) directly. The most common types of fuel cells in this category are molten carbonate fuel cells (MCFC), with an operating temperature range of 600–700°C, and solid oxide fuel cells (SOFC), which operate in the temperature range of 600–1000°C.

Other types of fuel cells can use non-hydrogen fuels directly, without the need for internal or external reforming. This category includes Direct Methanol Fuel Cells (DMFC) and Direct Carbon Fuel Cells (DCFC). The DMFC, also referred to as a Direct Alcohol Fuel Cell, which is mainly used in low-power portable applications, uses alcohol as its fuel and is considered to be a low-temperature fuel cell. On the other hand, the DCFC uses carbon as the fuel directly in the anode. Due to the thermodynamics of reactions taking place in a DCFC, it will usually result in a high-efficiency conversion. Table 4.1 gives a comparative summary of the structure, characteristics, benefits, and applications of different kinds of fuel cells.

Among different types of fuel cells, PEMFC is an attractive candidate for many applications such as vehicular applications and stationary and backup

TABLE 4.1

Comparison of Different Types of Fuel Cells

	AFC	PAFC	PEMFC	MCFC	SOFC
Electrodes	Transition metals	Carbon	Carbon	Nickel and nickel oxide	Metal cement
Catalyst	Platinum based	Platinum based	Platinum based	Non-noble metal	Non-noble metal
Electrolyte	Potassium hydroxide	Liquid phosphoric acid	Solid polymer	Liquid molten carbonate	Dense Yttria-stabilized Zirconia (Ceramic)
Operating temperature range (°C)	80–260	~200	50–80	600–700	600–1000
Charge (ion) carrier	OH^-	H^+	H^+	CO_3^{2-}	O^{2-}
Product water management	Evaporative	Evaporative	Evaporative	Gaseous product	Gaseous product
Product heat management	Process gas	Process gas + liquid cooling medium or steam generation	Process gas + liquid cooling	Internal reforming + process gas medium	Internal reforming + process gas
Need for internal reforming	No	No	No	Yes	Yes
CO tolerance	No, poison (< 50 ppm)	No, poison (< 1%)	No, poison (< 50 ppm)	Yes (fuel)	Yes (fuel)
Electrical efficiency (%)	~50	~40	40–50	45–55	50–60
Power density range (mW/ cm²)	150–400	150–300	300–1000	100–300	250–350
Power range (kW)	1–100	50–1000	10^{-3}–1000	100–10^5	5–10^5
Applications	Space, stationary power generation	Stationary, dispersed power plants, and on-site cogeneration power plants	Portable, transportation, and stationary power generation	Stationary and combined heat and power generation	Portable, transportation, stationary, and combined heat and power generation

Source: Nehrir, M. H., and C. Wang, *Modeling and Control of Fuel Cells Distributed Generation Applications*, Wiley–IEEE Hoboken Press, New Jersey, 2009.

power generation. The most favorable reasons for this interest are quick start-up and high electrical efficiency, approximately 40–50%. Due to these advantages, this chapter will focus on PEM fuel cells.

Proton exchange membrane fuel cells use a solid polymer as the electrolyte which is usually made from fluorinated sulfonic acid polymer. This Teflon-like electrolyte acts as a conductor of protons and an insulator of electrons. At the anode, during an oxidation reaction, hydrogen molecules are broken into electrons and hydrogen protons with the help of the catalyst. The hydrogen protons travel across the membrane (electrolyte) to reach the cathode surface where they engage in a reduction reaction with electrons passing from the anode to the cathode through the external load, to produce water. The anode reaction, cathode reaction, and the overall PEMFC reaction, respectively, are represented by the following equations:

$$2H_2 \rightarrow 4H^+ + 4e^- \tag{4.1}$$

$$O_2 + 4H^+ + 4e^- \rightarrow 2H_2O \tag{4.2}$$

$$2H_2 + O_2 \rightarrow 2H_2O \tag{4.3}$$

A schematic diagram and chemical reactions for the PEMFC are shown in Figure 4.1.

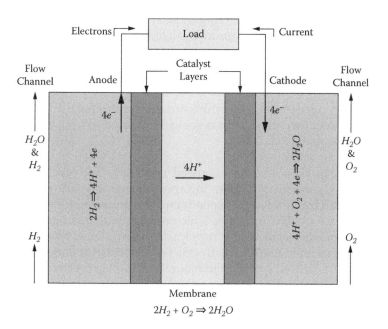

FIGURE 4.1
PEMFC schematic diagram and chemical reactions.

As mentioned earlier, due to the sensitivity of PEMFCs to high-temperature vapors, their operating temperature is limited to the range of 60–80°C. Humidity management is a very important issue in the efficient operation of PEMFCs. In most fuel cell systems, the by-product water is used to humidify the membrane; however, proper steps should be taken to prevent membrane flooding or dehydration. The flooding phenomenon occurs when the accumulation of generated water molecules in the membrane impedes the passing of hydrogen molecules, resulting in a decrease in the stack voltage. The generated hot water can also be used for purposes such as residential hot water needs. Furthermore, in Combined Heat and Power (CHP) applications, the exhaust heat is used for space heating or to operate a cogeneration system to generate more electricity. High-temperature fuel cells, like SOFCs, are more suitable for operating in CHP mode. In the CHP operation mode, SOFCs can reach efficiencies around 80%.

In the case of PEMFCs, a narrow range of low operating temperatures makes it difficult to use the rejected heat effectively. The need for pure hydrogen is another downside of PEMFCs because, as mentioned earlier, they are very sensitive to poisoning by impurities contained in the hydrogen such as ammonia, CO, and sulfur. Furthermore, if only hydrocarbon fuels are available, extensive fuel-processing systems will be required, negatively affecting system size, complexity, and cost, and resulting in efficiencies as low as 35% (Nehrir et al., 2009).

Despite the aforementioned obstacles, the PEMFC is the most developed fuel cell to date. It is already used in stationary power sources as backup generation and will soon be fully commercialized in automotive applications.

4.1.3 Typical Fuel Cell System Components

Auxiliary fuel cell components, which are responsible for the safe and efficient operation of fuel cells and managing their output power, are described below (Pukrushpan et al., 2004).

1. *Fuel Supply Subsystem:* As discussed previously, each fuel cell type requires a specific fuel. The incorporation of this subsystem into the overall fuel cell system depends on the required fuel. For example, in the case of PEM fuel cells, which require pure hydrogen, if pure hydrogen is not available, the supply subsystem is a fuel processor that converts its input fuel into hydrogen. For residential applications, supplying the fuel cell system with natural gas is often preferred because of its wide availability and extended distribution system.

2. *Reactant Flow Subsystem:* The reactant flow subsystem is comprised of all components that direct the inlet hydrogen and air into the fuel cell. These components may include pipes, inlet and outlet manifolds, pumps, motors, valves, pressure or flow regulators, and compressors

or blowers. The components in this subsystem are responsible for providing sufficient reactant flows to maintain the desired excess ratio, ensure fast and safe power transient responses, and minimize auxiliary power consumption.

3. *Thermal Management Subsystem:* The thermal management subsystem includes the fuel cell stack and the reactant temperature system's monitoring and control components. The thermal management subsystem is responsible for guaranteeing that the operating temperature of the system remains in an acceptable range while optimizing the power consumption in auxiliary fan and pumps. Temperature overshoot can result in phenomena such as membrane dehydration or physical damage to the cell structure, which, in turn, will seriously degrade the fuel cell performance. On the other hand, low operating temperatures will reduce the reaction rates and, therefore, efficiency of the fuel cell. As current is drawn from the fuel cell, heat is generated. For large stacks, like those required for vehicles, the heat generated cannot be passively dumped to the surrounding air by convection and radiation through the surface; rather, cooling systems are required to actively manage the heat produced. The thermal management of fuel cells is more difficult than that of internal combustion engines. First, the coolant typically used in the stacks is deionized water, which is not a very effective coolant. Second, the exhaust air exiting the low-temperature fuel cell stacks is also at a low temperature and, thus, is less able to carry out heat than the Internal Combustion Engines (ICE) exhaust gas, which is typically over 500°C.

4. *Water Management Subsystem:* The task of the water management subsystem is to maintain the membrane hydration at a desired value and to maintain a balance between system water usage and consumption. The main factors affecting membrane humidity are the reactant mass flow rate and the amount of water in the anode and cathode inlet flows. As current is depleted in the fuel cell, water molecules are both produced in the cathode and transferred from the anode to the cathode by means of hydrogen protons, which can affect the membrane humidity. Dry and flooded membranes are not favorable as they can cause high polarization losses. A water management subsystem monitors the membrane humidity and prevents these undesirable working conditions.

5. *Power Management Subsystem:* The power management subsystem is responsible for controlling the output power of the fuel cell stack. The current drawn from the fuel cell has a direct impact on other subsystems; however, without having some sort of control on the output current, it can be considered as a disturbance to the system.

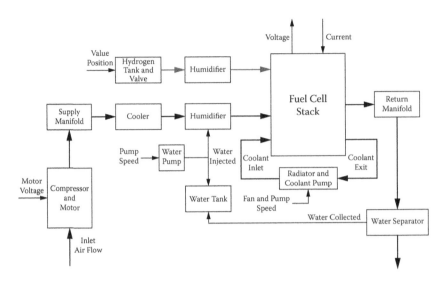

FIGURE 4.2
Typical fuel cell system block diagram. (From Pukrushpan, J. T., H. Peng, and A.G. Stefanopoulou, *ASME Journal of Dynamic Systems, Measurement, and Control*, 126, 14–25, 2004b.)

Power management becomes even more challenging when a secondary power source such as a battery is used along with the fuel cell. In such a configuration, the power management subsystem acts as a supervisory controller to provide a satisfactory transient response and achieve optimal overall system efficiency.

A block diagram of a fuel cell stack, including some of the ancillary components, is shown in Figure 4.2.

4.2 Fuel Cell System Model

There are many factors that impact the dynamics of PEM fuel cells such as material mass transport, reaction mechanisms, pressures, temperatures, and electrode concentrations that produce voltages and currents, in addition to overpotential and ohmic losses. The degree of complexity of a PEMFC model is dependent on the number of these factors that are considered. Some mechanistic and analytical models have been proposed to capture fuel cell dynamics in great detail (Rowe et al., 2001; Wang et al., 2004; Yao et al., 2004; Sousa et al., 2005). These studies account for the microscopic reaction dynamics and also structural analysis of the PEMFC. They are not practical to implement due to their high complexity and also the existence of unknown parameters in their structure, which are required for numerical simulation. In contrast, other works (Kim et al., 1995; Guzzella, 1999; Squadrito et al., 1999; Yerramalla et al., 2002) accounted for these dynamics by using more compact models that do not require complex solution techniques. They are semi-empirical models acquired from experimental data. These papers do not account for the fuel cell ancillary components, which are vital for the efficient and reliable operation of fuel cells.

Pukrushpan et al. (2002, 2004) used lumped-parameter, control-volume approaches to develop a dynamic model of a PEM fuel cell system. The model includes the compressor model, supply and return manifolds, air coolant, humidifier, anode and cathode flow, membrane hydration models, and a fuel cell polarization curve; however, they considered the temperature to be constant and neglected the charge double layer effect, which is a fast dynamic behavior that usually occurs near the electrode/electrolyte interfaces. In other words, the charged layer in the anode and cathode interfaces with electrolyte behaves as an electrical capacitor. More work has been conducted to include temperature and the charge double layer effect in PEM fuel cell models by Xue et al. (2004), Pathapati (2005), and Meyer et al. (2006). However, to be able to implement control methodologies on fuel cell systems, control-oriented models are needed (Grujicic et al., 2004). Control-oriented models consider the transient effects and neglect the spatial variations of model parameters. The dynamics of the electrochemical reactions and electrode electrical response are neglected due to their fast responses. However, dynamic behavior of other slower subsystems such as manifold filling, heat management, and the air compressor, as well as their interactions, are included in these models.

Recent work has been conducted in developing more detailed control-oriented PEMFC models. Gou et al. (2010) combined different models discussed earlier and also developed a nonlinear model to unify the steady-state and transient behavior of the polarization curve. However, they only considered the fuel cell itself and neglected the effect of the auxiliary components.

4.2.1 Polarization Curve Modeling

Fuel cell stack performance is typically characterized by its polarization curve, shown in Figure 4.3. The polarization curve can be separated into three different regions: activation overvoltage, ohmic overvoltage, and concentration overvoltage. Activation overvoltage corresponds to the energy barrier that must be overcome to initiate a chemical reaction between reactants. Activation overvoltage is a major source of voltage loss when the current density is small. It is not desirable to operate the fuel cell in this region due to its low efficiency. As the current density increases, the activation overvoltage becomes less significant while ohmic overvoltage becomes the dominant voltage loss. Ohmic overvoltage occurs due to resistive losses in the cell. These resistive losses occur within the electrolyte, the electrodes,

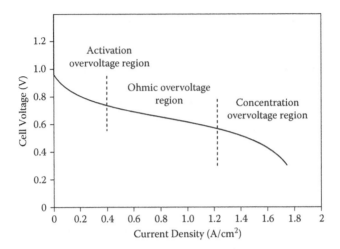

FIGURE 4.3
PEM fuel cell polarization curve.

and terminal connections in the cell. According to Ohm's law, the amount of voltage loss varies linearly throughout this region. As the current density approaches the limiting current density (i.e., the concentration of reactants at the catalyst surface reaches zero), concentration overvoltage becomes the dominant voltage loss. Concentration overvoltage is the voltage loss due to reactant starvation (i.e., reactants are consumed faster than they are supplied). The relationship between concentration overvoltage and current density is given by (Barbir, 2005)

$$v_{conc} = \frac{RT_{cell}}{2F} \ln\left(\frac{i_L}{i_L - i}\right) \tag{4.4}$$

where i_L is the limiting current density (A/cm²), i is the current density (A/cm²), R is the gas constant, T_{cell} is the cell temperature (K), and F is the Faraday constant (magnitude of electric charge per mole of electrons, i.e., 96,485 C/mol). As the current density approaches the limiting current density, the cell voltage drops rapidly. Since reactant starvation may cause permanent damage to the membrane, it is not desirable to operate the fuel cell in this region. The shape of the polarization curve mainly depends on the stack temperature and reactant pressure. Thus, the stack performance over its entire operation region can be described using a family of polarization curves measured for different operating conditions.

Different fuel cell voltage models have been used in the literature (Amphlett et al., 1995; Pukrushpan et al., 2004). A common fuel cell voltage model is

$$v_{fc} = E - v_{act} - v_{ohm} - v_{conc} \tag{4.5}$$

where E is the open circuit voltage and v_{act}, v_{ohm}, and v_{conc} are activation, ohmic, and concentration overvoltages, respectively. Concentration overvoltage is usually small under normal operation conditions (i.e., reactant starvation does not occur) and, thus, is neglected in some studies (Pathapati et al., 2005). The open circuit voltage is modeled using the Nernst equation (Amphlett et al., 1995; Pukrushpan et al., 2004),

$$E = 1.229 - 8.5 \times 10^{-4} \left(T_{cell} - 298.15\right) + 4.308 \times 10^{-5} T_{cell} \left(\ln\left(\frac{p_{an}^{H_2}}{p_{atm}}\right) + \frac{1}{2}\ln\left(\frac{p_{ca}^{O_2}}{p_{atm}}\right) \right)$$

(4.6)

where $p_{an}^{H_2}$, $p_{ca}^{O_2}$, and p_{atm} are hydrogen partial pressure (Pa), oxygen partial pressure (Pa), and atmospheric pressure (101 kPa), respectively. It can be observed from Figure 4.4 that fuel cell performance improves as reactant pressures increase, which is due to the fact that the chemical reaction is proportional to the partial pressures of the hydrogen and oxygen. However, higher pressure requires better sealing in the stack and additional compressor power. Increasing the stack temperature can also help increase the stack performance, as can be observed in Figure 4.5. However, the temperature should be below 100°C, which is the water boiling temperature. At this temperature, the water boils and the water vapor generated severely reduces the oxygen content, drastically reducing cell performance due to oxygen starvation. The activation overvoltage is modeled using the Tafel equation (Pukrushpan et al., 2004)

$$v_{act} = v_o + v_a \left(1 - e^{\xi_1 i}\right)$$

(4.7)

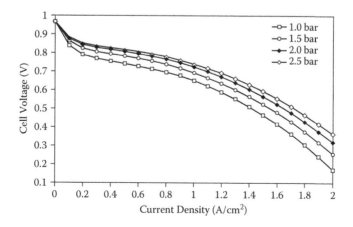

FIGURE 4.4
Effect of operating pressure on fuel cell polarization curve.

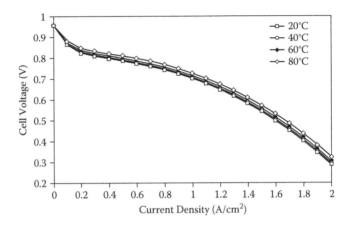

FIGURE 4.5
Effect of operating temperature on fuel cell polarization curve.

where v_o, v_a, and ξ_1 are empirical coefficients. The parameter v_o is

$$v_o = 0.27 - 8.5 \times 10^{-4} \left(T_{cell} - 298.15 \right) + 4.3085 \times 10^{-5}$$

$$\left[\ln \left(\frac{P_{ca} - p_{ca}^{sat}}{p_{atm}} \right) + \frac{1}{2} \ln \left(\frac{0.1173 \left(P_{ca} - p_{ca}^{sat} \right)}{p_{atm}} \right) \right] \tag{4.8}$$

and the parameter v_a is

$$v_a = \left(-1.618 \times 10^{-5} T_{cell} + 1.618 \times 10^{-2} \right) \left(\frac{p_{ca}^{O_2}}{0.1173 \times 10^5} + \frac{p_{ca}^{sat}}{10^5} \right)^2$$

$$+ \left(1.8 \times 10^{-4} T_{cell} - 0.166 \right) \left(\frac{p_{ca}^{O_2}}{0.1173 \times 10^5} + \frac{p_{ca}^{sat}}{10^5} \right) - 5.8 \times 10^{-4} T_{cell} + 0.5736 \tag{4.9}$$

The ohmic overvoltage is modeled using Ohm's law,

$$v_{ohm} = R_{ohm} I_{st} \tag{4.10}$$

where R_{ohm} is the cell resistance (Ω) and I_{st} is the stack current (A). The cell resistance depends on the membrane water content and stack temperature. It is empirically modeled by (Amphlett et al., 1995; Pathapati et al., 2005)

$$R_{ohm} = \xi_2 + \xi_3 T_{cell} + \xi_4 I_{st} \tag{4.11}$$

where ξ_2, ξ_3, and ξ_4 are parametric coefficients determined from experiments. The concentration overvoltage is (Pukrushpan et al., 2004)

$$v_{conc} = i\left(\xi_5 \frac{i}{i_{max}}\right)^{\xi_6} \tag{4.12}$$

where ξ_5, ξ_6, and i_{max} (A/cm^2) are empirical coefficients. The parameter ξ_5, which depends on stack temperature and reactant partial pressures, is

$$\xi_5 = \begin{cases} \left(7.16\times10^{-4}T_{cell} - 0.622\right)\left(\dfrac{p_{ca}^{O_2}}{0.1173\times10^5} + \dfrac{p_{ca}^{sat}}{10^5}\right) \\[2ex] +\left(-1.45\times10^{-3}T_{cell} + 1.68\right), \text{ if } \dfrac{p_{ca}^{O_2}}{0.1173} + p_{ca}^{sat} < 2p_{atm} \\[2ex] \left(8.66\times10^{-5}T_{cell} - 0.068\right)\left(\dfrac{p_{ca}^{O_2}}{0.1173\times10^5} + \dfrac{p_{ca}^{sat}}{10^5}\right) \\[2ex] +\left(-1.6\times10^{-3}T_{cell} + 0.54\right), \text{ if } \dfrac{p_{ca}^{O_2}}{0.1173} + p_{ca}^{sat} \geq 2p_{atm} \end{cases} \tag{4.13}$$

4.2.2 Reactant Supply System Model

A typical fuel cell reactant supply system is shown in Figure 4.6. The fuel cell air supply system consists of the following components: air compressor, humidifier, supply manifold, cathode, return manifold, and back pressure valve. The supply manifold inlet flow and return manifold outlet flow are regulated via the air compressor and a back pressure valve, respectively. The hydrogen supply system components include the following components: hydrogen tank, pressure regulator, and anode and purge valve.

4.2.2.1 Air Supply System Model

4.2.2.1.1 Compressor Model

Air compressors are typically used in PEM fuel cell systems for oxidant supply (Bao et al., 2006; Pukrushpan et al., 2004). The compressor flow, which depends on the compressor speed and the pressure ratio, is determined using a compressor map, as shown in Figure 4.7. The compressor map shows the compressor efficiency for different pressure ratios and mass flow factors. Lines for constant rotational speed factor are also plotted on the map. The mass flow factor and rotational speed factors are defined as $\frac{W_{cp}\sqrt{T_1}}{1000P_1}$ and $\frac{N}{\sqrt{T_1}}$,

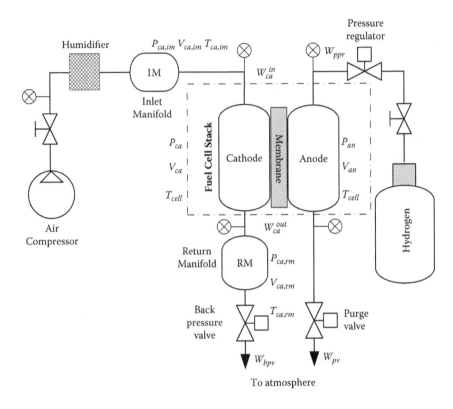

FIGURE 4.6
PEM fuel cell reactant supply system.

respectively. The parameter W_{cp} is the mass flow rate (g/s), T_1 is the compressor inlet temperature (293.15K), P_1 is the inlet pressure (i.e., atmospheric pressure), and N is the compressor rotor speed (rev/min).

The compressor is usually driven by an electric motor; therefore, the motor angular velocity is described by (Bao et al., 2006)

$$J_{eq}\dot{\omega}_m = T_m - f_{eq}\omega_m - \gamma T_{cp} \tag{4.14}$$

where ω_m is motor angular velocity (rad/s), γ is the gear ratio between the motor and compressor $(\gamma = \frac{\omega_{cp}}{\omega_m})$, J_{eq} is the combined inertia of the compressor and motor (kg·m²), T_m is the motor torque (N·m), f_{eq} is the friction coefficient (N·m/(rad/s)), and T_{cp} is the compressor torque (N·m). The motor torque is determined by the static model (Bao et al., 2006; Pukrushpan et al., 2004)

$$T_m = \frac{\eta_m c_t \left(k_m u_m - \frac{30}{\pi} c_e \omega_m\right)}{R_m} \tag{4.15}$$

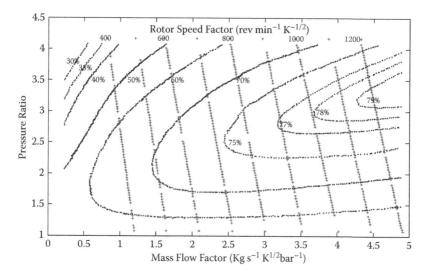

FIGURE 4.7
Compressor efficiency map. (From Larminie J., and A. Dicks, *Fuel Cell Systems Explained*, 2nd edition, John Wiley and Sons Ltd., Chichester, 2003.)

where u_m is the control input to the motor (0–10 V), η_m is the mechanical efficiency, and k_m, R_m, c_t, and c_e are motor constants. The compressor efficiency is

$$\eta_{cp} = g\left(\frac{p_{ca,im}}{p_{atm}}, \frac{W_{cp}}{10^3} \right) \tag{4.16}$$

where g is the compressor map, W_{cp} is the compressor flow rate (g/s), and $p_{ca,im}$ is the supply manifold pressure (Pa). The compressor map can be modeled using different methods—for example, back propagation neural network (Bao et al., 2006), Jensen and Kristensen nonlinear curve fitting method (Pukrushpan et al., 2004), and so on. The required compressor torque is determined by

$$T_{cp} = \frac{C_p T_{atm}}{10^3 \omega_{cp} \eta_{cp}} \left[\left(\frac{p_{ca,im}}{p_{atm}} \right)^{(\kappa-1)/\kappa} - 1 \right] W_{cp} \tag{4.17}$$

where C_p is the specific heat capacity of air (J·Kg⁻¹·K⁻¹), T_{atm} is the atmospheric temperature (293.15 K), and κ is the specific heat ratio.

4.2.2.1.2 Humidifier Model

Humidification is crucial for efficient fuel cell operation. If the relative humidity is too low, dehydration may occur, resulting in poor performance or even cell damage; if the relative humidity is too high, flooding may occur, resulting in an unacceptable voltage drop. External humidification is provided through humidifiers installed on the cathode and anode sides. The relative humidity regulated by the cathode side humidifier is

$$RH_{ca,hum} = \frac{p^v_{ca,hum}}{p^{sat}_{ca,hum}} \times 100\% \tag{4.18}$$

where $p^v_{ca,hum}$ (Pa) is the water vapor partial pressure and $p^{sat}_{ca,hum}$ (Pa) is the vapor saturation pressure, which is modeled by

$$p^{sat}_{ca,hum} = 10^3 \left(\frac{T_{ca,hum} - 273.15}{k} + 0.6105^b \right)^{\frac{1}{b}} \tag{4.19}$$

where $T_{ca,hum}$ is the humidifier temperature, and the parameters k and b are determined by nonlinear fitting. The water vapor partial pressure is

$$p^v_{ca,hum} = p_{ca,hum} \frac{n_v}{n} \tag{4.20}$$

where n_v is moles of the water vapor, n is the total moles of the gas mixture (i.e., oxygen, nitrogen, and water vapor), and $p_{ca,hum}$ (Pa) is the cathode humidifier pressure. Ignoring the pressure drop between the humidifier and the cathode inlet manifold, the cathode humidifier pressure is equal to the cathode inlet manifold pressure $p_{ca,im}$ (Pa). The water vapor mole fraction is

$$MF^v_{ca,hum} = \frac{n_v}{n} = \frac{p^{sat}_{ca,hum} RH_{ca,hum}}{p_{ca,hum}} \tag{4.21}$$

Assuming the mass flow rate at the humidifier inlet (dry air) is $W^{in}_{ca,hum}$ (g/s), which is equal to the compressor air flow rate W_{cp} (g/s), the water vapor mass flow rate and the mixture mass flow rate, respectively, in the cathode are

$$W^{out}_{ca,hum} = W^{in}_{ca,hum} + \frac{M_v MF^v_{ca,hum}}{M_a - M_a MF^v_{ca,hum}} W^{in}_{ca,hum}$$

$$W^v_{ca,hum} = \frac{M_v MF^v_{ca,hum}}{M_a - M_a MF^v_{ca,hum}} W^{in}_{ca,hum} \tag{4.22}$$

where M_a (g/mol) is the molar mass of air and M_v (g/mol) is the molar mass of water vapor (g/mol). The mass fractions of water vapor, oxygen, and nitrogen, respectively, are

$$\alpha_{ca,hum}^v = \frac{M_v MF_{ca,hum}^v}{M_a - M_a MF_{ca,hum}^v + M_v MF_{ca,hum}^v}$$

$$\alpha_{ca,hum}^{O_2} = \frac{0.233(M_a - M_a MF_{ca,hum}^v)}{M_a - M_a MF_{ca,hum}^v + M_v MF_{ca,hum}^v} \quad (4.23)$$

$$\alpha_{ca,hum}^{N_2} = \frac{0.767(M_a - M_a MF_{ca,hum}^v)}{M_a - M_a MF_{ca,hum}^v + M_v MF_{ca,hum}^v}$$

The molar mass of the gas mixture is

$$M_{ca,hum}^m = MF_{ca,hum}^v M_v + 0.21\left(1 - MF_{ca,hum}^v\right)M_{O_2} + 0.79\left(1 - MF_{ca,hum}^v\right)M_{N_2} \quad (4.24)$$

4.2.2.1.3 Cathode Inlet Manifold Model

The cathode inlet manifold pressure dynamics can be modeled using the ideal gas law:

$$\dot{P}_{ca,im} = \frac{R T_{ca,im}}{M_{ca,im}^m V_{ca,im}}\left(W_{ca,im}^{in} - W_{ca,im}^{out}\right) \quad (4.25)$$

where $P_{ca,im}$ is the cathode inlet manifold pressure (Pa), $T_{ca,im}$ is the cathode inlet manifold temperature (K), $M_{ca,im}^m$ is the molar mass of the gas mixture in the cathode inlet manifold (g/mol), $V_{ca,im}$ is the cathode inlet manifold volume (m^3), $W_{ca,im}^{in}$ is the inflow mass flow rate (g/s), $W_{ca,im}^{out}$ is the outflow mass flow rate (g/s), and R is the gas constant, that is, 8.314472 J/(K·mol). Since the cathode inlet manifold inflow is the same as the humidifier outflow

$$W_{ca,im}^{in} = W_{ca,hum}^{out}$$

$$M_{ca,im}^m = M_{ca,hum}^m \quad (4.26)$$

The outflow mass flow rate is (Bao et al., 2006)

$$W_{ca,im}^{out} = W_{ca}^{in} = k_{ca,in}(P_{ca,im} - P_{ca}) \quad (4.27)$$

where W_{ca}^{in} is the cathode inflow mass flow rate (g/s), $k_{ca,in}$ is the cathode inlet manifold flow coefficient (g/(Pa·s)), and P_{ca} is the cathode pressure (Pa).

4.2.2.1.4 Cathode Model

The oxygen consumption is (Pukrushpan et al., 2004)

$$CR_{O_2} = M_{O_2} \frac{n_{cell} I_{st}}{4F} \tag{4.28}$$

where CR_{O_2} is the oxygen consumption rate (g/s) and n_{cell} is the number of cells connected in series. The rate at which water vapor is generated at the cathode is (Pukrushpan et al., 2004)

$$GR_v = M_v \frac{n_{cell} I_{st}}{2F} \tag{4.29}$$

where GR_v is water vapor generation rate (g/s) and M_v is the molar mass of water vapor (g/mol). The cathode inlet flow can be written as

$$W_{ca}^{in} = W_{ca,in}^{out} = W_{ca}^{v,in} + W_{ca}^{O_2,in} + W_{ca}^{N_2,in}$$

$$= \alpha_{ca,hum}^v W_{ca}^{in} + \alpha_{ca,hum}^{O_2} W_{ca}^{in} + \alpha_{ca,hum}^{N_2} W_{ca}^{in} \tag{4.30}$$

The partial pressure derivatives for oxygen, nitrogen, and water vapor, respectively, are

$$\dot{p}_{ca}^{O_2} = \frac{RT_{cell}}{M_{O_2} V_{ca}} \left(W_{ca}^{O_2,in} - W_{ca}^{O_2,out} - CR_{O_2} \right)$$

$$\dot{p}_{ca}^{N_2} = \frac{RT_{cell}}{M_{N_2} V_{ca}} \left(W_{ca}^{N_2,in} - W_{ca}^{N_2,out} \right) \tag{4.31}$$

$$\dot{p}_{ca}^{v} = \frac{RT_{cell}}{M_v V_{ca}} \left(W_{ca}^{v,in} - W_{ca}^{v,out} + GR_v - W_{ca}^{v,phase} \right)$$

The total pressure in the fuel cell cathode side is

$$P_{ca} = p_{ca}^v + p_{ca}^{O_2} + p_{ca}^{N_2} \tag{4.32}$$

The mass flow rate out of the cathode is

$$W_{ca}^{out} = k_{ca,rm} (P_{ca} - P_{ca,rm}) \tag{4.33}$$

where P_{ca} and $P_{ca,rm}$ are fuel cell cathode pressure (Pa) and return manifold pressure (Pa), respectively, and $k_{ca,rm}$ is the flow coefficient (g/s/Pa). The mass flow rate out of the cathode can also be written as

$$W_{ca}^{out} = W_{ca}^{O_2,out} + W_{ca}^{N_2,out} + W_{ca}^{v,out} \tag{4.34}$$

where

$$W_{ca}^{O_2,out} = \frac{n_{ca}^{O_2} M_{O_2}}{n_{ca}^{O_2} M_{O_2} + n_{ca}^{N_2} M_{N_2} + n_{ca}^{v} M_v} W_{ca}^{out}$$

$$W_{ca}^{N_2,out} = \frac{n_{ca}^{N_2} M_{N_2}}{n_{ca}^{O_2} M_{O_2} + n_{ca}^{N_2} M_{N_2} + n_{ca}^{v} M_v} W_{ca}^{out} \qquad (4.35)$$

$$W_{ca}^{v,out} = \frac{n_{ca}^{v} M_v}{n_{ca}^{O_2} M_{O_2} + n_{ca}^{N_2} M_{N_2} + n_{ca}^{v} M_v} W_{ca}^{out}$$

and $n_{ca}^{O_2}$, $n_{ca}^{N_2}$, and n_{ca}^{v} are oxygen, nitrogen, and water vapor mole numbers, respectively, in the cathode.

4.2.2.1.5 Cathode Return Manifold Model

The partial pressure derivatives for the oxygen, nitrogen, and water vapor, respectively, in the return manifold are

$$\dot{p}_{ca,rm}^{O_2} = \frac{RT_{ca,rm}}{M_{O_2} V_{ca,rm}} \left(W_{ca,rm}^{O_2,in} - W_{ca,rm}^{O_2,out} \right)$$

$$\dot{p}_{ca,rm}^{N_2} = \frac{RT_{ca,rm}}{M_{N_2} V_{ca,rm}} \left(W_{ca,rm}^{N_2,in} - W_{ca,rm}^{N_2,out} \right) \qquad (4.36)$$

$$\dot{p}_{ca,rm}^{v} = \frac{RT_{ca,rm}}{M_v V_{ca,rm}} \left(W_{ca,rm}^{v,in} - W_{ca,rm}^{v,out} \right)$$

The total pressure in the return manifold is

$$P_{ca,rm} = p_{ca,rm}^{O_2} + p_{ca,rm}^{N_2} + p_{ca,rm}^{v} \qquad (4.37)$$

The mass flow rate out of the return manifold is

$$W_{ca,rm}^{out} = W_{ca,rm}^{O_2,out} + W_{ca,rm}^{N_2,out} + W_{ca,rm}^{v,out} \qquad (4.38)$$

where

$$W_{ca,rm}^{O_2,out} = \frac{n_{ca,rm}^{O_2} M_{O_2}}{n_{ca,rm}^{O_2} M_{O_2} + n_{ca,rm}^{N_2} M_{N_2} + n_{ca,rm}^{v} M_v} W_{ca,rm}^{out}$$

$$W_{ca,rm}^{N_2,out} = \frac{n_{ca,rm}^{N_2} M_{N_2}}{n_{ca,rm}^{O_2} M_{O_2} + n_{ca,rm}^{N_2} M_{N_2} + n_{ca,rm}^{v} M_v} W_{ca,rm}^{out} \qquad (4.39)$$

$$W_{ca,rm}^{v} = \frac{n_{ca,rm}^{v} M_v}{n_{ca,rm}^{O_2} M_{O_2} + n_{ca,rm}^{N_2} M_{N_2} + n_{ca,rm}^{v} M_v} W_{ca,rm}^{out}$$

and $n_{rm}^{O_2}$, $n_{rm}^{N_2}$, and n_{rm}^{v} are oxygen, nitrogen, and water vapor mole numbers, respectively, in the return manifold. The mass flow out of the return manifold is

$$W_{ca,rm}^{out} = W_{bpv} \tag{4.40}$$

where W_{bpv} (g/s) is the mass flow through the back pressure valve.

4.2.2.1.6 Cathode Back Pressure Valve Model

The back pressure valve at the return manifold exit is used to regulate the return manifold pressure, which can be modeled using the nozzle equation

$$W_{bpv} = \begin{cases} \dfrac{C_{bpv} A_{bpv} P_{ca,rm}}{\sqrt{RT_{ca,rm}}} \sqrt{\dfrac{2\kappa}{\kappa-1}\left[\left(\dfrac{P_{atm}}{P_{ca,rm}}\right)^{\frac{2}{\kappa}} - \left(\dfrac{P_{atm}}{P_{ca,rm}}\right)^{\frac{\kappa+1}{\kappa}}\right]} & \text{if} \quad \dfrac{P_{atm}}{P_{ca,rm}} > \left(\dfrac{2}{\kappa+1}\right)^{\frac{\kappa}{\kappa-1}} \\[4ex] \dfrac{C_{bpv} A_{bpv} P_{ca,rm}}{\sqrt{RT_{ca,rm}}} \kappa^{1/2}\left(\dfrac{2}{\kappa+1}\right)^{\frac{\kappa+1}{2(\kappa-1)}} & \text{if} \quad \dfrac{P_{atm}}{P_{ca,rm}} \leq \left(\dfrac{2}{\kappa+1}\right)^{\frac{\kappa}{\kappa-1}} \end{cases}$$

$$\tag{4.41}$$

where C_{bvp} is the back pressure valve discharge coefficient, A_{bvp} is the back pressure valve opening (m²), and $T_{ca,rm}$ is the cathode return manifold temperature (K).

4.2.2.2 Hydrogen Flow Model

The hydrogen flow system consists of a hydrogen tank, a proportional pressure regulator, an anode humidifier, an anode, and a purge valve. The proportional pressure regulator is used to control the flow from the hydrogen tank to minimize the pressure difference between the cathode and anode.

4.2.2.2.1 Proportional Pressure Regulator Model

The hydrogen flow rate though the proportional pressure regulator is (Pukrushpan et al., 2004)

$$W_{ppr} = k_1(P_{ca} - P_{an}) \tag{4.42}$$

where k_1 is a gain (g/s/Pa).

4.2.2.2.2 Anode Humidifier Model

Assuming the mass flow rate at the humidifier inlet (dry hydrogen) is $W_{an,hum}^{in}$, which is equal to the hydrogen flow through proportional pressure regulator

W_{ppr} the mass flow rate of the water vapor and the mixture mass flow rate, respectively, are

$$W_{an,hum}^{out} = W_{an,hum}^{in} + \frac{M_v MF_{an,hum}^v}{M_{H_2} - M_{H_2} MF_{an,hum}^v} W_{an,hum}^{in} \tag{4.43}$$

$$W_{an,hum}^v = \frac{M_v MF_{an,hum}^v}{M_{H_2} - M_{H_2} MF_{an,hum}^v} W_{an,hum}^{in} \tag{4.44}$$

where

$$MF_{an,hum}^v = \frac{p_{an,hum}^{sat} RH_{an,hum}}{p_{an,hum}} \tag{4.45}$$

and M_{H_2} (g/mol) is the hydrogen molar mass, M_v (g/mol) is the water vapor molar mass, $p_{an,hu}^{hum}$ is the water vapor saturation pressure at the anode humidifier, and $RH_{an,hum}$ is the desired relative humidity. The mass fractions of water vapor and hydrogen, respectively, are

$$\alpha_{an,hum}^v = \frac{M_v MF_{an,hum}^v}{M_{H_2} - M_{H_2} MF_{an,hum}^v + M_v MF_{an,hum}^v}$$

$$\alpha_{an,hum}^{H_2} = \frac{M_{H_2} - M_{H_2} MF_v}{M_{H_2} - M_{H_2} MF_v + M_v MF_v} \tag{4.46}$$

4.2.2.2.3 Anode Flow Model

The mass flow rate into the anode is

$$W_{an}^{in} = W_{an}^{H_2,in} + W_{an}^{v,in} = W_{an,hum}^{out} = \alpha_{an,hum}^{H_2} W_{an,hum}^{out} + \alpha_{an,hum}^v W_{an,hum}^{out} \tag{4.47}$$

where $W_{an}^{H_2,in}$ and $W_{an}^{v,in}$ are hydrogen and water vapor flow rates (g/s), respectively, into the anode. The rates of change of the mass of hydrogen and water vapor, respectively, in the anode are

$$\dot{m}_{an}^{H_2} = W_{an}^{H_2,in} - W_{an}^{H_2,purge} - W_{an}^{H_2,react}$$

$$\dot{m}_{an}^v = W_{an}^{v,in} - W_{an}^{v,purge} - W_{an}^{v,phase} \tag{4.48}$$

where $\dot{m}_{an}^{H_2}$ is the hydrogen mass change rate in the anode, which is determined by the hydrogen inflow rate, $W_{an}^{H_2,in}$, hydrogen outflow rate due to purging operations, $W_{an}^{H_2,purge}$, and hydrogen consumption rate, $W_{an}^{H_2,react}$, and $\dot{m}_{an}^v$ is the water vapor mass change rate in the anode, which is determined by the water vapor inflow rate, $W_{an}^{v,in}$, water vapor outflow rate due to purging operations, $W_{an}^{v,purge}$, and water vapor condensing rate, $W_{an}^{v,phase}$. It should be noted that the water transport due to the electro-osmotic drag (Pukrushpan et al., 2004) phenomenon (i.e., when water molecules are dragged along with the protons traveling through the membrane pores) is neglected here.

The hydrogen and water vapor partial pressure change rates, respectively, are

$$\dot{p}_{an}^{H_2} = \frac{RT_{cell}}{M_{H_2} V_{an}} \dot{m}_{an}^{H_2}$$

$$\dot{p}_{an}^{v} = \frac{RT_{cell}}{M_v V_{an}} \dot{m}_{an}^{v} \tag{4.49}$$

The hydrogen consumption is

$$W_{an}^{H_2, react} = M_{H_2} \frac{n_{cell} I_{st}}{2F} \tag{4.50}$$

The mass flow rate out of the anode is

$$W_{an}^{out} = W_{an}^{H_2, purge} + W_{an}^{v, purge} \tag{4.51}$$

where the mass flow rates of hydrogen and water vapor, respectively, due to the purge operation are

$$W_{an}^{H_2, purge} = \frac{n_{an}^{H_2} M_{H_2}}{n_{an}^{H_2} M_{H_2} + n_{an}^{v} M_v} W_{an}^{out}$$

$$W_{an}^{v, purge} = \frac{n_{an}^{v} M_v}{n_{an}^{H_2} M_{H_2} + n_{an}^{v} M_v} W_{an}^{out} \tag{4.52}$$

4.2.2.2.4 Purge Valve Model

The mass flow rate through the purge valve is modeled using the nozzle equation

$$W_{pv} = \begin{cases} \dfrac{C_{pv} A_{pv} P_{an}}{\sqrt{RT_{cell}}} \sqrt{\dfrac{2\kappa}{\kappa - 1} \left[\left(\dfrac{P_{atm}}{P_{an}} \right)^{\frac{2}{\kappa}} - \left(\dfrac{P_{atm}}{P_{an}} \right)^{\frac{\kappa+1}{\kappa}} \right]} & \text{if} \quad \dfrac{P_{atm}}{P_{an}} > \left(\dfrac{2}{\kappa + 1} \right)^{\frac{\kappa}{\kappa-1}} \\[4ex] \dfrac{C_{pv} A_{pv} P_{an}}{\sqrt{RT_{cell}}} \kappa^{1/2} \left(\dfrac{2}{\kappa + 1} \right)^{\frac{\kappa+1}{2(\kappa-1)}} & \text{if} \quad \dfrac{P_{atm}}{P_{an}} \le \left(\dfrac{2}{\kappa + 1} \right)^{\frac{\kappa}{\kappa-1}} \end{cases}$$

$$\tag{4.53}$$

where C_{pv} is the purge valve discharge coefficient, A_{pv} is the purge valve opening (m²), and P_{an} is the anode total pressure (Pa). The mass flow through the purge valve can also be written as

$$W_{pv} = W_{an}^{out} \tag{4.54}$$

4.2.3 Thermal Management System Model

Thermal management is an essential subsystem within a PEM fuel cell to ensure precise stack temperature regulation. Depending on the materials used in the fuel cell membrane, strict thermal limits must be satisfied to prevent membrane damage. Furthermore, accurately controlling relative humidity (RH) is of the utmost importance in the performance and durability of fuel cells. One way to maintain a proper RH is through controlling the internal temperature of the fuel cell.

Many studies have been conducted to describe the temperature dynamics inside the fuel cell stack (e.g., Golbert et al., 2004). The models used in these studies are typically spatial time-dependant partial differential equations, which are usually too complex for real-time implementation in thermal management systems. In most studies conducted in the area of fuel cells, the temperature dynamics inside the stack are usually ignored and a constant temperature inside the stack is assumed.

In McKay et al. (2008), a gas humidification apparatus was designed and constructed to control the humidity and temperature of fuel cell reactants. For this system, a low-order, control-oriented model of the humidification system thermal dynamics was developed and identified using experimental data. Hu et al. (2010) employed a complete coolant circuit modeling method, which includes a PEMFC thermal model, a water reservoir model, a pump model, a bypass valve model, a heat exchanger model, and the PEMFC electrochemical model.

The temperature management system introduced by (Kolodziej et al., 2007) and (Choe et al., 2008) will be presented in this subsection. It employs coolant flow to maintain a constant temperature inside the fuel cell stack. A thermal model based on a Continuous Stirred-Tank Reactor (CSTR) model is introduced, which results in a lumped parameter, nonlinear, first-order differential equation. The model is an energy balance of the coolant according to

$$\dot{E}_{stored} = \dot{E}_{in} - \dot{E}_{out} + \dot{E}_{gen} \tag{4.55}$$

where the energy terms $\dot{E}_{in}$, $\dot{E}_{out}$, $\dot{E}_{stored}$, and $\dot{E}_{gen}$ represent the input energy, output energy, stored energy, and generated energy inside the coolant, respectively

$$\dot{E}_{in} = \dot{m}_{in}\left(u + Pv + \frac{V^2}{2} + gZ\right)_{in} + \dot{q}_{in} \tag{4.56}$$

$$\dot{E}_{out} = \dot{m}_{out}\left(u + Pv + \frac{V^2}{2} + gZ\right)_{out} + \dot{w}_{out} \tag{4.57}$$

$$\dot{E}_{gen} = 0 \tag{4.58}$$

$$\dot{E}_{stored} = \frac{d}{dt}(\rho_c V_c C_{p,c} T_{stk,out}) \tag{4.59}$$

and $\dot{m}_{in}$ is the inlet coolant mass flow rate (g/s), $\dot{m}_{out}$ is the outlet coolant mass flow rate (g/s), ρ_c is the coolant density (g/m³), $C_{p,c}$ is the coolant specific heat (J/g°C), V_c is the coolant effective volume in the stack (m³), u is the coolant internal energy (J), P is the coolant pressure (Pa), v is the coolant specific volume (m³/g), $V^2/2$ is the coolant kinetic energy (J), gZ is the coolant potential energy (J), $\dot{q}_{in}$ is the heat transfer to the fuel cell (W), and $\dot{w}_{out}$ is the work done by the coolant (W). It is assumed that the height and velocity changes between the coolant inlet and outlet are negligible; therefore, the sum of potential and kinetic energy of the inlet and outlet coolant is equal. It is also worth noting that V_c is not the actual volume of the coolant within the fuel cell stack. Kolodziej et al. (2007) refer to it as the "effective" volume of the coolant, which artificially accounts for the stack's thermal mass. The value of V_c, determined empirically using an optimization algorithm introduced in (Kolodziej et al., 2007), is $V_c = 7.5 \times 10^{-4}\ m^3$.

It is assumed that the only source of power loss in the stack is heat generation which is directly transferred to the coolant. In other words, a portion of the output power that does not generate its nominal voltage produces heat. Furthermore, the stack temperature is assumed to be equal to the coolant outlet temperature, $T_{stk,out}$. Figure 4.8 shows a schematic of the fuel cell stack thermal control volume. The heat generation inside the coolant and the work done by it (i.e., $\dot{w}_{out}$) are assumed to be negligible. In this case, Equation 4.55 can be written as

$$\frac{d}{dt}(\rho_c V_c C_{p,c} T_{stk,out}) = \dot{m}_{in}(u + Pv)_{in} - \dot{m}_{out}(u + Pv)_{out} + \dot{q}_{in} \qquad (4.60)$$

Coolant mass flow into and out of the stack remains constant, $\dot{m}_{in} = \dot{m}_{out} = \dot{m}_c$. Also, since the potential and kinetic energies of the inlet and outlet coolant

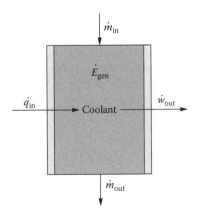

FIGURE 4.8
Fuel cell stack thermal control volume. (From Kolodziej, J. R., *Journal of Fuel Cell Science and Technology*, 4, 255–260, 2007.)

are assumed to be equal, the terms corresponding to these energies cancel. The enthalpy of a process is the summation of the internal energy and mechanical work, that is, $h = u + Pv$. Furthermore, if the temperature dependency of specific heat C_p is negligible, the enthalpy will be $h = C_p T$. Based on these assumptions and considering the fact that $(\rho_c V_c C_{p,c})$ is a time-invariant scalar, Equation 4.60 further reduces to

$$(\rho_c V_c C_{p,c})\dot{T}_{stk,out} = \dot{m}_c C_{p,c}(T_{stk,in} - T_{stk,out}) + \dot{q}_{in} \tag{4.61}$$

The heat transfer to the coolant, which is the energy loss due to the efficiency of the fuel cell, is

$$\dot{q}_{in} = (v_{ideal} - v_{actual})I_{st} = (v_{nom} - v_{ave})n_{cell}I_{st} \tag{4.62}$$

where v_{nom} is the nominal single cell voltage in a PEM fuel cell at 25°C (V), which is calculated from the free enthalpy of reaction. The voltage v_{ave} is the measured average cell voltage of the entire stack (V). The variables v_{ideal} and v_{actual} are the stack nominal and measured voltages (V), respectively. Equation 4.62 determines the value for the energy transferred to the coolant, which is assumed to be measurable.

4.2.4 Simulation Studies

4.2.4.1 Fuel Cell Simulations

In this section, fuel cell dynamic responses (i.e., open circuit stack voltage, power, and reactant partial pressures) are studied for constant compressor motor control voltages and cathode back pressure valve openings using step changes in the current demand. The simulation studies are conducted in MATLAB® using the parameters listed in Table 4.2. Anode pressure is controlled using a proportional pressure regulation valve to minimize the pressure difference between the cathode and anode. The simulation results are shown in Figures 4.9–4.12. Figure 4.9 shows the responses of the fuel cell stack voltage, V_{st}, power output P_{st}, and compressor flow rate W_{cp}. It can be observed that the steady-state stack voltage drops greatly as the current increases, which is due to two causes. First, since a fixed compressor control voltage is used, the partial pressures of the reactants drop as the current increases, resulting in a smaller open circuit voltage. The other is the increased ohmic overvoltage, which is proportional to the stack current. The cell voltage drop due to low oxygen concentration is usually avoided by adjusting the compressor control voltage to maintain a suitable constant oxygen excess ratio, which is defined by

$$\lambda_{O_2} = \frac{W_{ca}^{O_2,in}}{CR_{O_2}} \tag{4.63}$$

TABLE 4.2

Simulation Parameters

Symbol	Value	Description
A (cm^2)	1000	Cell active area
b	0.145	Water vapor saturation pressure model coefficient
C_{bpv}	1.264×10^{-2}	Back pressure valve discharge coefficient
$c_e \left(\dfrac{V}{rad/s} \right)$	6.49×10^{-2}	Motor electromotive coefficient
$C_p \left(\dfrac{J}{kg \cdot K} \right)$	1004	Air specific heat capacity
$c_t \left(\dfrac{N \cdot m}{A} \right)$	0.62	Motor torque coefficient
$f_{eq} \left(\dfrac{N \cdot m}{rad/s} \right)$	0	Equivalent friction coefficient
$i_{max} \left(\dfrac{A}{cm^2} \right)$	2.2	Polarization curve model coefficient
J_{eq} (kg $\cdot$ m^2)	2.0	Equivalent inertia
k	98.27	Water vapor saturation pressure model coefficient
$k_{ca,in}$ (g/s/Pa)	1.3672×10^{-2}	Cathode inlet flow coefficient
$k_{ca,in}$ (g/s/Pa)	1.2524×10^{-2}	Cathode outlet flow coefficient
k_1 (g/s/Pa)	0.021	Proportional pressure regulator gain
k_m	30	Motor constant
n_{cell}	540	Number of cells
$R \left(\dfrac{J}{K \cdot mol} \right)$	8.314472	Gas constant
R_m (Ω)	0.159	Motor internal resistance
$T_{ca,hum}$ (K)	303.15	Cathode humidifier temperature
T_{cell} (K)	348.15	Fuel cell stack temperature
V_{an}(m^3)	9.6×10^{-3}	Anode volume
$V_{ca,im}$(m^3)	7×10^{-3}	Cathode inlet manifold volume
V_{ca}(m^3)	10.8×10^{-3}	Cathode volume
$V_{ca,rm}$(m^3)	2.4×10^{-3}	Cathode return manifold volume
v_o(V)	Equation 1.8	Polarization curve model coefficient
v_a(V)	Equation 1.9	Polarization curve model coefficient
ξ_1	10	Polarization curve model coefficient
ξ_2	1.605×10^{-5}	Polarization curve model coefficient
ξ_3	-3.5×10^{-5}	Polarization curve model coefficient
ξ_4	8×10^{-5}	Polarization curve model coefficient
ξ_5	Equation 1.13	Polarization curve model coefficient

TABLE 4.2 (CONTINUED)

Simulation Parameters

Symbol	Value	Description
ξ_6	2.0	Polarization curve model coefficient
γ	5	Gear ratio between motor and compressor
η_m	0.98	Motor shaft mechanical efficiency
κ	1.4	Air specific heat ratio

A high oxygen excess ratio corresponds to a high oxygen partial pressure, which increases the cell voltage. However, more power is required to drive the air compressor. A typical value of oxygen excess ratio is 2.0 (Pukrushpan et al., 2004). It is also noted that the compressor flow rate increases with the stack current, which is due to the lowered pressure ratio, as can be observed in Figure 4.10.

Figure 4.11 shows the responses of partial pressures of different gas species including oxygen, hydrogen, nitrogen, and cathode and anode water vapors. The oxygen partial pressure drops as the current increases, causing a larger oxygen consumption rate. The same trend is also observed with the hydrogen partial pressure. The cathode and anode water vapors do not change since they have reached the saturation point.

Figure 4.12 shows the simulation results of pressure variations at different fuel cell locations. It can be observed that all of the pressures on the fuel cell cathode side share the same pattern, which is reasonable considering that constant control signals are used. The cathode and anode total pressures drop as the current increases due to increased oxygen and hydrogen consumption, respectively.

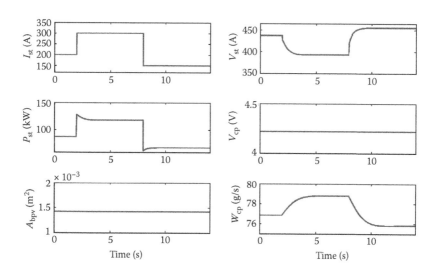

FIGURE 4.9

Fuel cell system simulation results for current step changes with constant compressor voltage and fixed cathode back pressure valve opening.

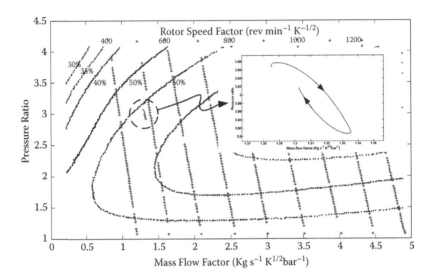

FIGURE 4.10
Effect of stepping current on compressor operating points.

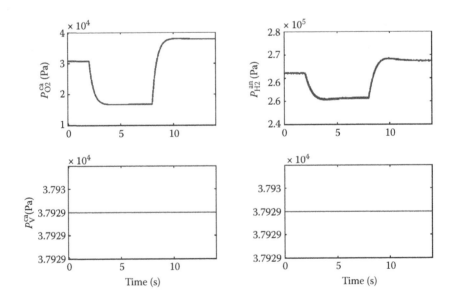

FIGURE 4.11
Simulated reactant partial pressures for input given in Figure 4.9.

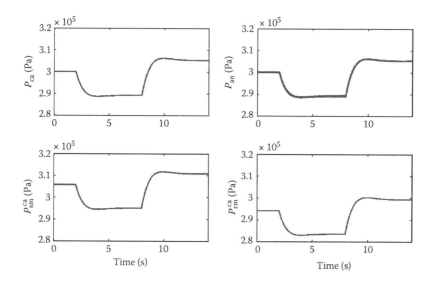

FIGURE 4.12
Effect of step changes in current on fuel cell total pressures.

The time required for the pressures to reach the steady state depends on the compressor dynamics. The anode pressure is controlled by a proportional pressure regulator to track the cathode pressure, as can be observed from the top two subplots in Figure 4.12. One disadvantage with a constant compressor control voltage is that the compressor flow cannot be regulated according to the fuel cell load status, as shown in Figure 4.13. The air compressor reference flow is calculated according to the current and the oxygen excess ratio. As the current increases, the compressor flow is significantly lower than the reference flow, resulting in oxygen starvation and possible membrane damage. As the current decreases, the compressor flow is significantly higher than the reference flow, which deteriorates the total fuel cell energy efficiency.

4.2.4.2 Thermal Management Simulations

In this section, the open loop response of the thermal management system, introduced in Section 4.2.3, to a current step changes is simulated. It is assumed that a pump is used to generate the coolant flow rate, which has first order dynamics. Therefore, the transfer function between the pump output mass flow rate and its input voltage is

$$\frac{\dot{m}_c(s)}{V(s)} = \frac{k_c}{\tau_c s + 1} \tag{4.64}$$

where the values of the pump time constant and gain, respectively, are $\tau_c = 2s$ and $k_c = 30$ g/V. The fuel cell stack output current profile used to simulate the system is shown in Figure 4.14.

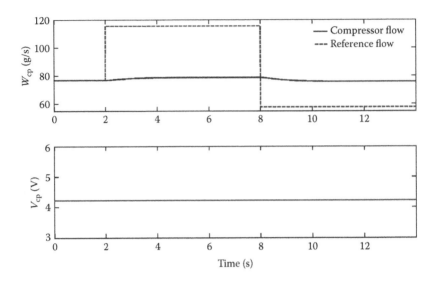

FIGURE 4.13
Open loop compressor flow with constant control voltage and step changes in current.

The nominal and average voltages of each cell are 1.23 V and 0.8 V, respectively. Therefore, the heat loss generated due to the difference between the nominal and actual stack voltage in Equation 4.61 is

$$\dot{q}_{in} = (1.23 - 0.8) n_{cell} I_{st} \tag{4.65}$$

In practice, the input temperature of the coolant is not constant. Therefore, in the simulations the coolant input temperature is assumed to vary sinusoidally

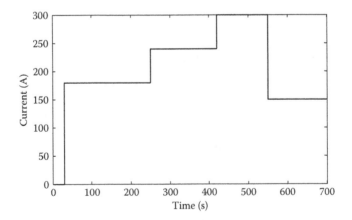

FIGURE 4.14
Fuel cell stack current profile used for simulation studies.

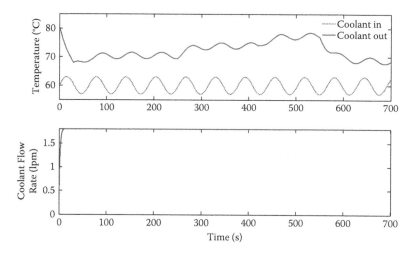

FIGURE 4.15
Effect of small coolant mass flow rate on stack temperature for current step changes.

around a constant temperature of 60°C with an amplitude of 3°C and frequency of 1.6×10^{-2} Hz.

In these simulations, a constant voltage is applied to the pump and its effect on the stack temperature, which is assumed to be the coolant outlet temperature, is investigated. Figure 4.15 shows the inlet temperature, outlet temperature, and mass flow rate of the coolant for a pump voltage of 1 V. The effect of applying a voltage with amplitude of 9 V to the pump is also shown in Figure 4.16. As expected, it can be seen that as the coolant mass flow rate increases, the stack temperature decreases.

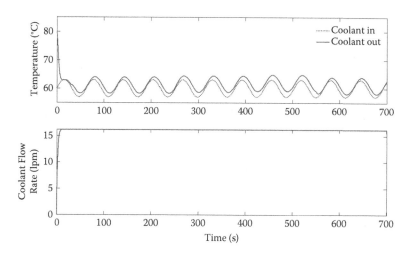

FIGURE 4.16
Effect of large coolant mass flow rate on stack temperature for current step changes.

4.3 Fuel Cell System Control

Achieving efficient PEM fuel cell performance poses a challenging problem for control system design and implementation due to the high degree of interactions between different subsystems. Control algorithms are required for the regulation of air and fuel flow to provide sufficient partial pressures of the reactants, pressure regulation to prevent membrane damage due to the difference in pressures between the reactants, thermal management to maintain the desired temperature, and power management to ensure the demanded power is delivered. Another issue is water management. In order for the fuel cell to work properly, its Membrane Electrode Assembly (MEA) should have sufficient moisture to prevent dehydration. However, too much water prevents reactant delivery, thus decreasing fuel cell efficiency.

In PEM fuel cell systems, the inlet cathode and anode pressures should be the same. Bao et al. (2006) implemented an observer-based LQG controller and a nonlinear neural network controller to regulate the pressure difference. Yu et al. (2006) have employed a proportional plus integral (PI) controller to regulate the stack pressure (which is assumed to be equivalent to cathode pressure) and cathode flow rate.

The control unit should ensure that oxygen starvation does not occur to guarantee the safe operation of the PEM fuel cell. In Suh et al. (2006), a decentralized controller was developed to minimize oxygen starvation by controlling the air flow. They also included a DC/DC controller to regulate the output voltage. To determine the air flow set-point, Tekin et al. (2006) employed a fuzzy supervision method.

Pukrushpan et al. (2004) introduced the oxygen excess ratio as the ratio of the supplied oxygen to oxygen used in the cathode to ensure sufficient oxygen supply and to achieve maximum power. They also showed that an oxygen excess ratio of 2 to 4 ensures maximum power output based on the stack current. To control the oxygen excess ratio, Grujicic et al. (2004) implemented a feedforward controller, and Pukrushpan et al. (2004, 2002) combined a feedforward controller with an LQR observer-based feedback controller. Danzer et al. (2008) employed a nonlinear model-based controller with an observer for this purpose. The observer presented in these papers is used to estimate the oxygen partial pressure in the cathode. Furthermore, it can be used to detect both temporary shortages of oxygen and peak cathode pressure. Pressure regulation is also achieved with a fast proportional feedback controller in these studies.

Another control objective is to regulate the output voltage while accounting for the safe operation of the fuel cell. This objective is accomplished in Na et al. (2007) by using an exact linearization based nonlinear controller by manipulating the stack current; however, in practical applications, stack current is usually considered as a disturbance to the system. Yang et al. (2007) applied an adaptive control strategy to adjust the air flow rate while keeping the load current and hydrogen flow rate constant in the presence of plant variations. A two-input two-output model was proposed in Wang et al. (2008), with the inputs being the inlet air and fuel flow rates, and the outputs being stack current and voltage. In this work, robust control algorithms were implemented to design multivariable H_∞ controllers to regulate the output voltage by manipulating the air and hydrogen flow rates.

Maintaining the desired stack temperature in an acceptable range is a very important issue, especially in high-power fuel cells. Based on the models in Equations 4.61 and 4.62, Choe et al. (2008) used linearization, based on a Taylor series expansion, to design a PI controller to regulate the temperature. However, using almost the same models, Kolodziej et al. (2007) developed feedback linearization to design a feedforward plus feedback controller that adjusts the coolant mass flow rate to regulate the stack temperature at a desired value. Hu et al. (2010) employed an incremental fuzzy controller with integral action based on their complete coolant circuit model to keep the PEMFC temperature within a desired temperature range.

4.3.1 Flow Control

4.3.1.1 Cathode Flow and Pressure Control

Cathode flow and pressure are usually controlled via the air compressor and back pressure valve, respectively. The desired flow rate is calculated according to the oxygen consumption rate and a desired oxygen excess ratio λ_{O_2} (typically 2). In this subsection, two PI controllers are developed to control the air compressor and back pressure valve. A PI controller can be described by

$$u(t) = k_p e(t) + k_i \int_0^t e(t)dt \tag{4.66}$$

where k_p and k_i are the proportional and integral gains, respectively, and $e(t)$ is the error. Dynamics of the back pressure valve opening can be modeled using a first order system

$$\frac{A_{bpv}(s)}{u_{bpv}(s)} = \frac{k_{bpv}}{\tau_{bpv}s + 1} \tag{4.67}$$

where u_{bpv}, A_{bpv}, k_{bpv}, and τ_{bpv} are the control signal (V), valve opening (m²), gain (m²/V), and time constant (s), respectively. To regulate the cathode pressure, a PI controller is developed and the control signal is described by

$$u_{bpv}(t) = k_p^{bpv} e_{p_{ca}} + k_i^{bpv} \int_0^t e_{p_{ca}}(t) \tag{4.68}$$

where k_p^{bpv} and k_i^{bpv} are controller gains and $e_{p_{ca}}$ is the cathode pressure error.

$$e_{p_{ca}} = P_{ca}^r - P_{ca} \tag{4.69}$$

where the cathode reference pressure is P_{ca}^r, which is set to 3×10^5 Pa. The air compressor is also controlled by a PI controller given by

$$u_{cp}(t) = k_p^{cp} e_{W_{cp}} + k_i^{cp} \int_0^t e_{W_{cp}}(t) \qquad (4.70)$$

where k_p^{cp} and k_i^{cp} are controller gains and $e_{W_{cp}}$ is the compressor flow error

$$e_{W_{cp}} = W_{cp}^r - W_{cp} \qquad (4.71)$$

where the reference flow rate, W_{cp}^r, is calculated from the oxygen excess ratio, λ_{O_2}, and oxygen consumption rate, CR_{O_2}, using

$$W_{cp}^r = \lambda_{O_2} \frac{CR_{O_2}}{MF_{O_2}^{air}} \qquad (4.72)$$

where $MF_{O_2}^{air}$ is the oxygen mass fraction in air (approximately 0.233). A series of current step tests is used as a simulation scenario and the results are shown in Figures 4.17–4.20. The compressor flow controller parameters are $k_p^{cp} = 2 \times 10^{-2}$ (V/(kg/s)) and $k_i^{cp} = 4 \times 10^{-2}$ (V/kg). The cathode back pressure

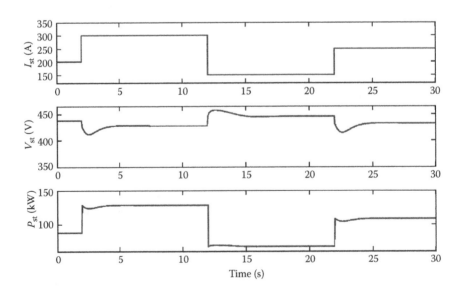

FIGURE 4.17
Stack current, voltage, and power with regulated air flow and cathode pressure.

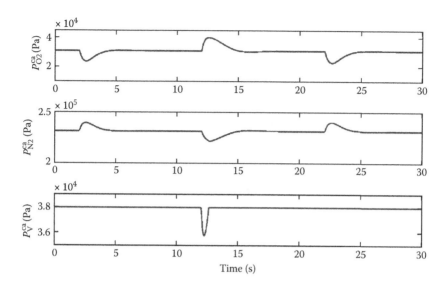

FIGURE 4.18
Cathode gas species partial pressures for current step changes.

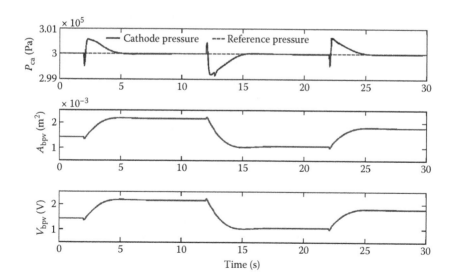

FIGURE 4.19
Cathode pressure and back pressure valve opening for current step changes.

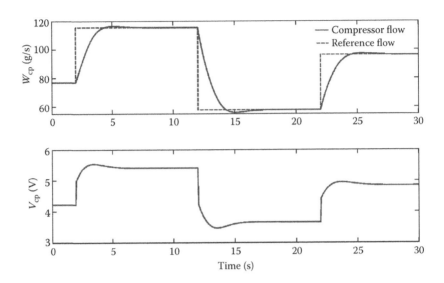

FIGURE 4.20
Compressor air flow and control voltage for current step changes.

valve controller parameters are $k_p^{bvp} = 1 \times 10^{-4}$ (V/(kg/s)) and $k_i^{bvp} = 1 \times 10^{-3}$ (V/kg). It can be observed that the stack voltage drops as the current increases during the steady state, which is due to the increase of activation, ohmic, and concentration overvoltages. The voltage drop is more significant during the transient period due to the decrease of reactant partial pressures, which results in a smaller cell open circuit voltage.

The partial pressures of oxygen, nitrogen, and water vapor for the simulation scenario are shown in Figure 4.18. It can be observed that the oxygen partial pressure decreases as the current increases due to the increasing oxygen consumption and reaches a steady value as the compressor flow reaches a steady state. The water vapor partial pressure drops as the current decreases, which is due to reduced water vapor generation at low currents.

The cathode pressure is regulated using a back pressure valve, as shown in Figure 4.19. The cathode pressure is maintained at 3.0×10^5 Pa during the steady state. The cathode pressure initially drops as the current increases, which is due to the increase in oxygen consumption rate.

The performance of the compressor air flow controller is shown in Figure 4.20. It can be seen that the actual compressor flow tracks the reference very well. However, due to the inertia of the motor and compressor rotor, the air flow cannot be changed instantaneously, which results in a significant decrease in the cell voltage during the transient period, as shown in Figure 4.17.

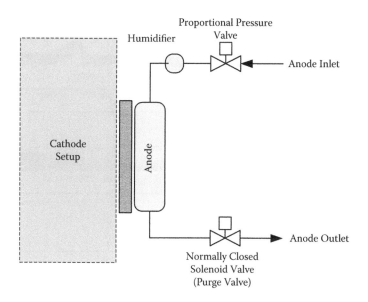

FIGURE 4.21
Dead-end anode configuration with purge valve.

4.3.1.2 Anode Pressure Control

Anode gas species typically include hydrogen and water vapor. To reduce the pressure difference between the anode and cathode, the anode pressure is controlled by a proportional pressure regulator. The flow going through the proportional pressure regulator is

$$W_{ppr}^{an} = k_1(P_{ca} - P_{an})$$ (4.73)

where k_1 is the proportional gain (g/s/Pa). The anode setup of the fuel cell, as shown in Figure 4.21, consists of a pressure regulator, humidifier, fuel cell anode, and purge valve. The fuel cell is operated with a dead-end anode configuration, which means the anode outlet is normally closed. It is more efficient than an open-end configuration and much less complex than a recirculation configuration. However, it may generate excessive water that blocks the hydrogen channel and, thus, it is necessary to periodically purge water by opening the purge valve. The purge valve can be implemented with a normally closed type solenoid valve at the exit of the hydrogen channel. The anode simulation results for the current step test above are shown in Figures 4.22 and 4.23. As can be observed in Figure 4.22, the purge valve opens every 5 seconds for a duration of 10 ms, which induces spikes in both hydrogen flow and anode pressure since the

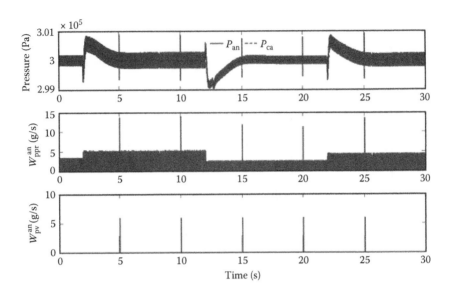

FIGURE 4.22
Anode total pressure and flow rates with purge operation for current step changes.

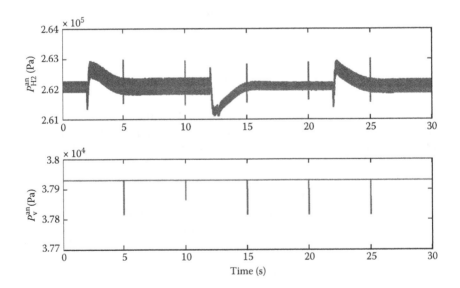

FIGURE 4.23
Anode hydrogen and water vapor partial pressures with purge operation for current step changes.

proportional pressure valve has to compensate for the purge flow. It is also noted that the anode pressure tracks the cathode pressure very well. The small oscillations in the anode pressure are due to the hydrogen consumption and proportional control method employed. It can also be observed in Figure 4.23 that the water vapor drops below the saturation point when the purge valve opens.

4.3.2 Temperature Control

Using coolant systems installed around the fuel cell stack to control stack temperature is one of the most common techniques used for thermal management. The coolants typically employed are water and ethylene glycol–based water solutions. The control objective is to guarantee that the stack temperature stays in an acceptable range. Another important consideration that should be taken into account when designing a thermal management system is the minimization of pump power consumption to increase the overall system efficiency. In this subsection, a control strategy is proposed to maintain the stack temperature in an acceptable range using water as the coolant. To this end, a feedback linearization controller introduced by Kolodziej et al. (2007) is designed based on Equation 4.61 to change the coolant mass flow rate such that the desired temperature is achieved. Furthermore, a PI controller is used to regulate the pump, which generates the desired mass flow rate. Figure 4.24 shows the block diagram of the system with the aforementioned controllers.

The assumptions that have been made in the simulation of the thermal management system in this section are the same as the assumptions in Section 4.2.4.2. The control goal of the thermal management system is to regulate the coolant outlet temperature at 70°C. The feedback linearization controller used here has the following structure:

$$\dot{m}_c = \frac{K(rVC_P)_c(T_{sp} - T_{stk,out}) - \dot{q}_{in}}{C_{P,c}(T_{stk,in} - T_{stk,out})} \tag{4.74}$$

where the control gain K is chosen by trial and error as $K = 0.9$ to achieve the fastest transient response with the lowest possible overshoot. Finally, the

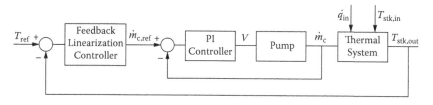

FIGURE 4.24
Block diagram of the fuel cell thermal management system.

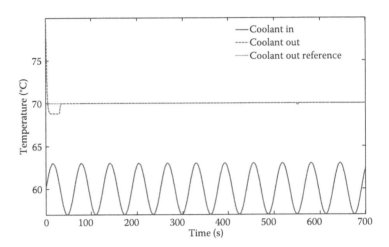

FIGURE 4.25
Coolant out, coolant in, and reference temperature for current step changes.

proportional and integral gains of the PI controller are calculated such that the closed-loop characteristic polynomial system has a damping ratio of 0.8 and natural frequency of 10 Hz. The values of these gains are $K_p = 0.5121$ and $K_I = 0.3145$. Figure 4.25 shows the coolant reference, input, and output temperatures. The pump output flow rate and its input voltage are shown in Figures 4.26 and 4.27, respectively.

As seen from the simulation results, this method can be implemented to guarantee that the stack temperature stays in the acceptable range. The coolant

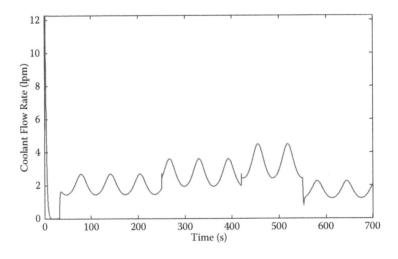

FIGURE 4.26
Coolant flow rate for current step changes.

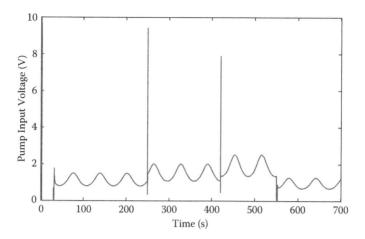

FIGURE 4.27
Pump input voltage for current step changes.

inlet temperature and the energy that is transferred to the coolant are considered as measurable disturbances to the system. The employed feedback linearization controller rejects these disturbances regardless of their signal form. Therefore, this thermal management system will still be effective for different scenarios of coolant inlet temperature and power losses. Furthermore, the control methodology is designed such that it ensures that the input voltage to the pump does not exceed its nominal operating limits while providing enough coolant mass flow rate to regulate the stack temperature.

4.4 Summary, Conclusions, and Future Work

As a novel power source, fuel cells are a promising candidate for transportation, stationary power generation, and portable applications due to their high efficiency and power density, and low operating temperature and emissions. This chapter reviewed the current progress in dynamic control-oriented modeling and control of hydrogen fuel cells. A polarization curve was first developed using electrochemical equations and empirical models. The effects of fuel cell temperature and reactant partial pressure were then evaluated. It is demonstrated that fuel cell performance can be improved by increasing temperature and/or reactant pressures. A comprehensive fuel cell model incorporating manifold filling dynamics, water vapor phase change, electrochemical reactions, and thermal management was developed. Open loop simulation studies were conducted. It was shown that proper controls are required to avoid oxygen starvation and improve efficiency. Two controllers were then designed for air compressor flow control and cathode pressure control via an air compressor and a back pressure valve installed at the exit of cathode return manifold. Simulation studies demonstrated that these two controllers are capable of regulating the compressor flow rate and cathode pressure simultaneously. The anode pressure was regulated using a proportional pressure valve designed to track the cathode pressure so that the pressure difference between the cathode and anode is minimized. To avoid flooding in the hydrogen channel, a simulation including a purge valve operating periodically at the hydrogen channel exit was conducted. The simulation studies showed that the proportional pressure valve can track the cathode pressure very well. To maintain a specific fuel cell working temperature, a fuel cell thermal management system was developed. The fuel cell temperature was regulated by adjusting the coolant flow rate. The simulation results indicated that the thermal control system can successfully regulate the fuel cell temperature.

There are several future directions for the work outlined in this chapter. In regard to modeling, the fuel cell model developed in this chapter could be further improved by incorporating the water transport between cathode and anode sides due to the electro-osmotic drag and back diffusion phenomenon. More experimental work could be conducted to verify models and systematic techniques could be developed to efficiently compute empirical parameters. During the past decade, substantial research has been dedicated to fuel cell simulation and control studies. However, due to the complex and nonlinear nature of the fuel cell system, specialized control strategies dedicated to a certain objective (e.g., pressure control, flow rate control, temperature regulation) are typically developed. For a typical fuel cell system, however, an integrated control system capable of achieving high-level objectives, such as maximum fuel efficiency, minimum fuel consumption, oxygen starvation avoidance, quick start-up, and so on, is still lacking. To improve fuel cell performance, systematic control systems that regulate the entire system still need to be developed. Future control systems will either be robust to the inherent uncertainties in the fuel cell system or be adaptable to changes in the system by adjusting the model and, hence, controller parameters in real time. Another future direction is the dynamic modeling and control of hybrid systems (i.e., fuel cells integrated with other energy sources such as internal combustion engines, advanced battery systems, etc.).

References

Amphlett, J., R. Baumert, R. Mann, B. Peppley, and P. Roberge. 1995. Performance modeling of the Ballard Mark IV solid polymer electrolyte fuel cell. *Journal of Electrochemical Society*, Vol. 142, pp. 9–15.

Bao, C., M. Ouyang, and B. Yi. 2006. Modeling and control of air stream and hydrogen flow with recirculation in a PEM fuel cell system—I. control-oriented modeling. *International Journal of Hydrogen Energy*, Vol. 31, pp. 1879–1896.

Barbir, F. 2005. *PEM fuel cells—theory and practice*. Elsevier Academic Press.

Choe, S., J. Ahn, J. Lee, and S. Baek. 2008. Dynamic simulator for a PEM fuel cell system with a PWM DC/DC converter. *IEEE Transactions on Energy Conversion*, Vol. 23, pp. 669–680.

Danzer, M.A., J. Wilhelm, H. Aschemann, and E.P. Hofer. 2007. Model-based control of cathode pressure and oxygen excess ratio of a PEM fuel cell system. *Journal of Power Sources*, Vol. 176, pp. 515–522.

Golbert, J., and D.R. Lewin. 2004. Model-based control of fuel cells: (1) regulatory control. *Journal of Power Sources*, Vol. 135, pp. 135–151.

Gou, B., W. Na, and B. Diong. 2010. *Fuel cells: modelling, control, and applications*. Boca Raton, CRC Press.

Grujicic, M., K.M. Chittajallu, E.H. Law, and J.T. Pukrushpan. 2004. Model-based control strategies in the dynamic interaction of air supply and fuel cell. *Journal of Power and Energy*, Vol. 218, pp. 487–499.

Guzzella, L. 1999. Control oriented modeling of fuel cell based vehicles. NSF Workshop on the Integration of Modeling and Control for Automotive Systems, June 5–6, Santa Barbara, California.

Hu, P., G. Cao, X. Zhu, and M. Hu. 2010. Coolant circuit modeling and temperature fuzzy control of proton exchange membrane fuel cells. *International Journal of Hydrogen Energy*, Vol. 35, pp. 9110–9123.

Kim, J., S. Lee, S. Srinivasan, and C. Chamberlin. 1995. Modeling of proton exchange membrane fuel cell performance with an empirical equation. *Journal of the Electrochemical Society*, Vol. 142, pp. 2670–2674.

Kolodziej, J.R., 2007. Thermal dynamic modeling and nonlinear control of a proton exchange membrane fuel cell stack. *Journal of Fuel Cell Science and Technology*, Vol. 4, pp. 255–260.

Larminie, J., and A. Dicks. 2003. *Fuel cell systems explained*, 2nd edition. Chichester, John Wiley and Sons Ltd.

McKay, D.A., A.G. Stefanopoulou, and J. Cook. 2008. Model and experimental validation of a controllable membrane-type humidifier for fuel cell applications. American Control Conference, pp. 312–317, June 11–13, Seattle, Washington.

Meyer, R.T., and B. Yao. 2006. Modeling and simulation of a modern PEM fuel cell system. *Fuel Cell Science, Engineering and Technology*, pp. 133–155.

Na, W., and B. Gou. 2007. Exact linearization based nonlinear control of PEM fuel cells. IEEE Power Engineering Society General Meeting, pp. 1–6, June 24–28, Tampa Bay, Florida.

Nehrir, M. H., and C. Wang. 2009. *Modeling and control of fuel cells distributed generation applications*. Hoboken, New Jersey: Wiley–IEEE Press.

Pathapati, P., X. Xue, and J. Tang. 2005. A new dynamic model for predicting transient phenomena in a PEM fuel cell system. *Renewable Energy*, Vol. 30, pp. 1–22.

Pukrushpan, J.T., A.G. Stefanopoulou, and H. Peng. 2002. Modeling and control for PEM fuel cell stack system. *American Control Conference*, Vol. 4, pp. 3117–3122, May 8–10, Anchorage, Alaska.

Pukrushpan, J. T., A.G. Stefanopoulou, and H. Peng. 2004a. *Control of fuel cell power systems: Principles, modeling, analysis, and feedback design.* London, New York: Springer.

Pukrushpan, J.T., H. Peng, and A.G. Stefanopoulou. 2004b. Control-oriented modeling and analysis for automotive fuel cell systems. *ASME Journal of Dynamic Systems, Measurement and Control*, Vol. 126, pp. 14–25.

Rowe, A., and X. Li. 2001. Mathematical modeling of proton exchange membrane fuel cells. *Journal of Power Sources*, Vol. 102, pp. 82–96.

Sousa, R., and E.R. Gonzalez. 2005. Mathematical modeling of polymer electrolyte fuel cells. *Journal of Power Sources*, Vol. 147, pp. 32–45.

Squadrito, G., G. Maggio, E. Passalacqua, F. Lufrano, and A. Patti. 1999. Empirical equation for polymer electrolyte fuel cell (PEFC) behaviour. *Journal of Applied Electrochemistry*, Vol. 29, pp. 1449–1455.

Suh, K.-W., and A.G. Stefanpolou. 2006. Effects of control strategy and calibration on hybridization level and fuel economy in fuel cell hybrid electric vehicle. SAE Technical Paper Series, SAE 2006-01-0038.

Tekin, M., D. Hissel, M.-C. Pera, and J.-M. Kauffmann. 2006. Energy consumption reduction of a PEM fuel cell motor-compressor group thanks to efficient control laws. *Journal of Power Sources*, Vol. 156, pp. 57–63.

Wang, C. 2004. Fundamental models for fuel cell engineering. *Chemical Reviews*, Vol. 104, pp. 4727–4765.

Wang, F., H. Chen, Y. Yang, and J. Yen. 2008. Multivariable robust control of a proton exchange membrane fuel cell system. *Journal of Power Sources*, Vol. 177, pp. 6118–6123.

Xue, X., J. Tang, A. Smirnova, R. England, and N. Sammes. 2004. System level lumped-parameter dynamic modeling of PEM fuel cell. *Journal of Power Sources*, Vol. 133, pp. 188–204.

Yang, Y., F. Wang, H. Chang, Y. Ma, and B. Weng. 2006. Low power proton exchange membrane fuel cell system identification and adaptive control. *Journal of Power Sources*, Vol. 164, pp. 761–771.

Yao, K., K. Karan, K. McAuley, P. Oosthuizen, B. Peppley, and T. Xie. 2004. A review of mathematical models for hydrogen and direct methanol polymer electrolyte membrane fuel cells. *Fuel Cells*, Vol. 4, pp. 3–29.

Yerramalla, S., A. Davari, and A. Feliachi. 2002. Dynamic modeling and analysis of polymer electrolyte fuel cell. Power Engineering Society Summer Meeting, Vol. 1, pp. 82–86, July 25–25, Chicago, Illinois.

Yu, Q., A.K. Srivastava, S.Y. Chloe, and W. Gao. 2006. Improved modeling and control of a PEM fuel cell power system for vehicles. Southeast Conferennce, pp. 331–336, March 31–April 2, Memphis, Tennessee.

Section II

Market Transformation and Applications

5

Market Transformation Lessons for Hydrogen from the Early History of the Manufactured Gas Industry

Marc W. Melaina

CONTENTS

5.1 Introduction: Lessons from the History of Manufactured Gas

The manufactured gas industry supplied high-quality urban lighting and heating for more than a century, from the early 1820s until the 1950s. Gas lighting dominated the high end of the lighting market, in parallel with relatively inexpensive whale oil and kerosene lighting, and eventually lost market share to even higher-quality electric lighting. The history of the gas light era offers lessons on quality energy services being provided through networked energy infrastructures and therefore contains interesting analogies to hydrogen infrastructure. This history is a mixture of successes and failures, with lessons on how technological innovations, market competition, adaptability, and intentional market transformation efforts can combine to shift the course of large energy systems. This chapter reviews this history with the goal of informing hydrogen infrastructure development challenges. The manufactured gas history is also rich with lessons for a broader range of technology policy issues, as demonstrated by several in-depth studies (Stotz and Jamison 1938; Castaneda 1999; Tarr 1999).

From 1825 to about 1890, gas lighting systems were established in every major city in the United States as well as in thousands of smaller cities, factories, and industrial facilities. The superior quality and convenience of gas streetlamps and household gas lights were a remarkable improvement over traditional whale oil lamps and tallow candles. Gas streetlamps allowed downtown commercial districts to thrive during evening hours, they illuminated great theaters, libraries, and statehouses, and they bolstered industrial productivity by extending working hours in factories. In addition, manufactured gas pipeline systems were some of the first large, networked energy industries to be planned and governed by way of contracts negotiated between government and private enterprise. The decline of the manufactured gas industry after 1925 also provides lessons on quality energy services, supply and demand dynamics, and technological change. Electricity eventually dominated the lighting market at high volumes, and improved welding techniques allowed cheap natural gas to be piped long distances to the gates of major U.S. cities.

In his historical review of the introduction of energy technologies during the Industrial Revolution, Nye notes that most energy technologies or services began as luxury or novelty items and then evolved, sometimes quickly, into basic necessities (1999). This same pattern can be seen in the history of public gas streetlamps, and to a lesser degree in private household gas lighting. Manufactured gas was primarily used by wealthy households that could afford high-quality lighting and were willing to pay a significant premium over whale oil or kerosene lamps. In contrast, gas street lighting became a fundamental public service in nearly all American and European cities and was financed through long-term contracts with municipal governments. Public street lamps had burned whale oil before the introduction of gas lighting. Historian Wolfgang Schivelbusch explains that street whale oil lamps were not intended to "illuminate the street but to serve as navigation beacons, guiding travelers much like today's runway lights guide airplane pilots" (Schivelbusch 1988, cited in Baldwin 2004, 751). Even with whale oil lamps in place, large cities were dangerous places at night, and few residents ventured out after dark. The introduction of gas streetlamps created safe urban zones with greatly reduced crime rates. The historical geographer Mark Bouman observes that "gas lamps were seen as powerful tools in suppressing immorality and crime and as symbols of progress" (Bouman 1987, cited in Baldwin 2004, 750). Reliance on gas streetlamps became deeply entrenched, as was apparent from the high level of concern expressed for public safety (sometimes bordering on hysteria) in the event of blackouts caused by fires or labor strikes at gas plant facilities (Baldwin 2004).

Interestingly, the public–private contractual agreements that procured this new public service also resulted in a physical and financial foundation from which gas companies could expand into the much more lucrative household lighting market. Municipal governments allowed gas companies to tear up city streets to lay pipelines and established long-term franchises to help finance the large capital outlays. At the time wealthy residential neighborhoods were located near city centers, so pipeline networks could expand to reach these households at low cost. This resulted in the classic monopoly conditions of large upfront costs, declining costs with increased production, and large barriers to entry for competitors. Extensive pipeline networks were developed within each company's franchise territory, and mergers and takeovers became common as demand increased and technology improved. These pipeline networks were eventually converted over to natural gas as transmission lines reached cities. At first natural gas was a supplement, and later manufactured gas plants were shut down.

The sections below explore these issues in more detail by delving deeper into the history of the manufactured gas industry, drawing comparisons and contrasts, and highlighting potentially valuable analogies and lessons. Section 5.2 examines various side-by-side comparisons between the two energy systems, including physical and chemical properties, costs, production processes, and system configurations. Section 5.3 examines infrastructure developments over time, reviewing five major phases in the history of manufactured gas. Finally, Section 5.4 concludes with five key analogies or lessons for hydrogen based upon this historical review.

5.2 Side-by-Side Comparisons

Manufactured gas and hydrogen infrastructure systems both rely upon a gaseous "energy carrier" as an intermediary between primary energy resources and end-use applications. Electricity serves a similar function as an energy carrier, and all three systems have required (or would require, in the case of a large-scale hydrogen infrastructure) extensive transmission, storage, and distribution systems linking remote production plants and end-use locations. Other energy carrier systems, such as natural gas-to-liquids, coal-to-liquids, methanol, dimethylether (DME) or ammonia systems, would also involve extensive production or delivery infrastructure development if employed on a large scale. Energy carrier systems are distinct from the direct use of biomass or fossil fuels resources, which can serve as their own energy carriers (e.g., burning wood, coal, petroleum products, or natural gas directly). By comparing manufactured gas and hydrogen systems side by side, in terms of physical properties, costs, production methods, and systems configurations, the similarities between the two fuels relative to other energy carrier systems become apparent. Four somewhat general similarities between the two fuels are reviewed below.

1. Compared to other fuel types, manufactured gas and hydrogen share common physical and chemical properties. Both fuels are gaseous under ambient conditions and both tend to be transmitted in pipelines and stored in large pressurized containers (though hydrogen may be transmitted or stored economically by other means and requires pipeline steels that resist embrittlement under pressure). Several types of manufactured gas production processes were used over time, but typical "town gas" was usually about 50% hydrogen. Both fuels have low volumetric energy density compared to natural gas, and therefore transmission, storage, and distribution systems have relatively large capital costs and large economies of scale, resulting in declining marginal costs with increased production.

2. Both fuels can be made from multiple primary energy resources. This is true for hydrogen more so than for manufactured gas, but the latter had been produced from whale oil, biomass resources (such as wood, pitch, and even animal fat from slaughterhouses), coal, and various fractions of crude oil. This diversity in feedstock type, in addition to a diversity of production methods, introduces a degree of complexity into the product chain. The combination of multiple, and sometimes competing, production methods with common systems of transmission, storage, and distribution results in complicated regulatory approaches and market responses.

3. Both fuels can be and have been used in multiple end-use applications. The early manufactured gas industry predominantly focused on illumination, but competition and adaptation shifted the industry toward household heating and cooking and eventually integration with the steel industry. Hydrogen is currently used in petroleum refining, petrochemical processing, and fertilizer production, with smaller volumes used in electronics and food processing. Future hydrogen end uses will likely focus on transportation and stationary

fuel cell applications, though demand for use in current applications will continue.

4. Over its long history, the manufactured gas industry adapted in response to multiple technological innovations and shifting market conditions and future hydrogen systems will likely adapt to analogous internal and external influences. Technological innovations in both upstream fuel systems and end-use technologies tended to improve the competitiveness of the manufactured gas industry over time, but eventually market conditions favored electric lighting and manufactured gas production technologies merged with the steel industry as long-distance natural gas pipelines began supplying gas to the city gate of municipal gas distribution systems. Shifts of a similar scale and type may also face future hydrogen systems and would likely depend upon both technological innovations (e.g., fuels cell performance, hydrogen storage systems) and changing market conditions (e.g., the price of oil and natural gas, climate change regulations). This being said, rates of technological change are faster today than in the 1800s.

Figure 5.1 provides an overview of the major infrastructure components of past manufactured and natural gas systems, as well as near-term hydrogen

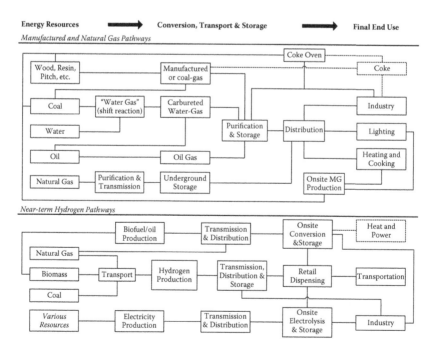

FIGURE 5.1
Major manufactured, natural gas, and near-term hydrogen pathways. (Credit: M. Melaina, NREL.)

infrastructure systems. Various types of energy resources are indicated on the left-hand side of the figure. These resources are converted, transported, and stored before reaching a final end-use application, indicated on the right-hand side of the figure. Manufactured gas systems, shown in the top half of the figure, were initially based upon the conversion of various biomass products, and later the conversion of coal and oil products. These production systems were eventually displaced when natural gas was introduced via long-distance pipelines between 1920 and 1960.

Three major manufactured gas production processes were introduced in succession: carbonization of coal to produce coal gas, dissociation of steam to produce water gas, and hydrocracking of petroleum products to produce either pure oil gas or a mixture of oil gas and water gas referred to at the time as carbureted water gas. Following the production process, manufactured gas was purified, stored, and distributed for various end uses, primarily for lighting during the first 100 years of the industry and later for residential heating and cooking and industrial applications. In addition to large-scale urban distribution systems, manufactured gas was produced on-site for various smaller applications, such as individual buildings, factories, or industrial applications. Gas production eventually shifted to the steel industry with the rise of electric lighting and natural gas. These trends and other technological changes are reviewed chronologically in Section 5.3.

The bottom half of Figure 5.1 provides an overview of comparable near-term hydrogen infrastructure components. Though hydrogen can be produced from any primary energy source, early large-scale systems will likely to focus on the thermochemical conversion of natural gas, coal, and biomass resources. Electrolysis on-site (i.e., near the point of use) is another near-term pathway, indicated at the bottom of the figure as drawing upon various types of primary energy resources. In the pathways shown, the resources on the left-hand side of the figure would be delivered to a central, "city-gate," or on-site location before being converted to hydrogen. In some applications, heat or electricity could also be produced.

5.2.1 Gaseous Fuel Compositions and Relative Household Costs

Table 5.1 compares typical compositions and heating values for various gaseous fuels. The fuels are indicated in their approximate order of chronological introduction, as discussed in more detail in Section 5.3. Coal gas consists of predominantly hydrogen and methane and is produced via carbonization: the heating of coal in retorts and subsequent collection of the gases released. Coal carbonization was the dominant method of producing manufactured gas during the early decades of the industry, though some systems also produced gas from wood, pitch, rosin, whale oil, and even animal fat from slaughterhouses (Clark 1963, p. 21; Tarr 2004; Waples 2005, p. 30). Water gas was produced by exposing steam to incandescent coal, resulting in a gas rich in carbon monoxide and with a relatively low heating value. Carbureted

TABLE 5.1

Typical Compositions and Heating Values for Various Gaseous Fuels

Gas Attribute	Symbol/ Units	Coal Gas	Water Gas	Carbureted Water Gas	Town Gas	Natural Gas	Hydrogen
Gas Composition							
Hydrogen	H_2	53%	49%	32%	48%	—	100%
Carbon Monoxide	CO	5%	40%	40%	17%	—	—
Methane	CH_4	33%	1%	19%	24%	93%	—
Other Hydrocarbons	C_nH_m	3%	0%	13%	5%	4%	—
Carbon Dioxide	CO_2	1%	5%	2%	2%	—	—
Nitrogen	N_2	5%	5%	3%	5%	—	—
Oxygen	O_2	—	0.5%	0.4%	0.2%	—	—
Lower Heating Value	Btu/scf	517	273	572	479	918	275

Notes: All percentages are percent volumes. Coal gas or "retort coal," water gas or "blue gas," and carbureted water gas from Speight, J. G. and M. Dekker, *The Chemistry and Technology of Coal*, Speight, New York, 1994, Table 1.

water gas was produced by thermally cracking light oil products within a water gas production plant to increase the heating value and improve luminosity. Section 5.2.2 discusses these production processes in greater detail. Town gas was the dominant manufactured gas product during the second half of the industry's history, and resulted from a mixture of coal gas and carbureted water gas. Natural gas and hydrogen are also shown for comparison.

Rough estimates of capital investments and infrastructure extent exist within the historical literature. These estimates can be used to compare the relative cost and scale of hydrogen and manufactured gas industries. The *American Gas-Light Journal* reported that 183 gaslight companies in 1859 consisted of a total investment of $31.7 million (Castaneda 1999, p. 52). Stotz and Jamison (1938, p. 4) report 297 companies (62% more than the *American Gas-Light Journal* estimate) serving 5 million persons by 1859, with a total capitalization of $42 million (32% greater than the *American Gas-Light Journal* estimate). With most cities having adopted gas lighting during or after the 1930s (see Figure 5.8), these estimates suggest a total capital investment of approximately $0.8 to $1 billion (in 2007 dollars) over a 20- to 30-year period (though inflating dollars of such a long time period is highly speculative[*]). For a direct comparison to hydrogen infrastructure development, and assuming mobile refueler or on-site steam methane reforming hydrogen stations costing $2–3 million each, an equivalent investment today would establish approximately 300–400 hydrogen stations. With reference to the aggressive scenario presented in Greene et al. (2008), this number of stations would provide about 25–30% of the coverage needed to satisfy the refueling needs of

---

[*] All 2007 dollar comparisons are based upon the historical inflation indices prepared by Rober Sahr, available online at http://oregonstate.edu/cla/polisci/faculty-research/sahr/sahr.htm.

early adopter markets in large urban areas within the 2016–2019 time period, several years after the initial mass production of fuel cell vehicles.

As another point of reference, the manufactured gas industry produced at least 25,000 million scf of gas in 1890 (EPA 1985), which is the energy equivalent of the hydrogen needed to support 200,000 hydrogen vehicles.* During the 1920s, this volume would have increased by more than 10-fold to approximately 300 billion scf, or the energy equivalent of the hydrogen needed to support about 2.5 million hydrogen vehicles.

Comparisons can also be made on a typical household expenditure basis. A typical household supplied with manufactured gas between 1815 and 1855 was relatively wealthy and spent approximately $22 per year on gas lighting (Stotz and Jamison 1938, p. 7), or about $500 per year when converted to 2007 dollars.[†] As a reference, the average household spent $535 on electricity for lighting and small appliances in 2001.[‡] Contemporary households spend much more on transportation fuels: The average 2001 household spent $1520 on gasoline.[§] Wealthy households during the early introductory period of manufactured gas therefore spent about as much on gas lighting as an average household in 2001 spent on electricity for lighting and small appliances, and about one-third as much as the average household in 2001 spent on gasoline.

5.2.2 Manufactured Gas Production Processes

The manufactured gas industry shifted production methods over time in response to both technology improvements and innovations and market conditions for different energy resources. For example, early systems relied upon wood, pitch, or rosin, but then nearly all large systems switched to coal as coal became both more available (as a result of railroad expansion) and relatively inexpensive. Later, water gas became a major production method as that technology became more mature. Shortly after the discovery of oil, the production of oil gas through a thermal cracking process became a major production method and benefited from the low cost of lighter oil fractions. Each of these three major production processes, resulting in coal gas, water gas, and carbureted water gas, is reviewed in the following sections.

5.2.2.1 Carbonization

Carbonization is the process of heating coal in retort reactors to drive off volatile constituents, which were captured as product gas, leaving behind

---

* Assuming 12,000 miles per year and 65 miles per kg.
† See Kemp, Schot, et al. (1998).
‡ Does not include space heating, electric-air cooling, water heating, or refrigerators. Derived from EIA's 2001 Residential Energy Consumption Survey, available online: http://www.eia.doe.gov/emeu/recs/recs2001/ce_pdf/enduse/ce1-3e_hhincome2001.pdf.
§ EIA Household Vehicles Energy Use data, online at http://www.eia.doe.gov/emeu/rtecs/nhts_survey/2001/tablefiles/page_a02.html\Wealthy.

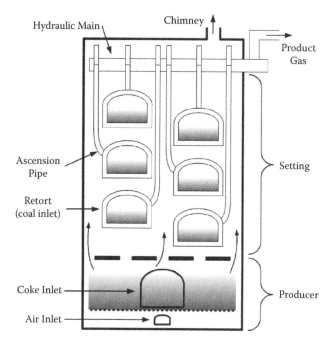

FIGURE 5.2
A bed of six horizontal retorts. (Credit: M. Melaina, NREL.)

primarily a coke by-product. Tars and other impurities are then removed from the gas before storage and distribution.* The first retorts were horizontal retorts, varying in length between about 8 to 20 feet and typically 1.5 to 2 feet across. These retorts were round, oval or "D" shaped—domed with flat bottoms. Retorts were encased in a brick "setting" that contained the hot producer gas used to heat the retorts. Figure 5.2 shows the configuration of a "bed" of six retorts in a setting, with a producer oven (or simply "producer") located below the setting. Producers are fueled with coke, which is combusted in a limited oxygen atmosphere through the controlled introduction of air, resulting in a product gas consisting primarily of carbon monoxide and nitrogen. The carbon monoxide then combusts with additional air supplied to the setting, further increasing the temperature of the setting and retorts. Combustion chamber temperatures were kept at about 1350°C (Smith 1945, p. 33), and the settings were designed to increase the free circulation of combustion gases and ensure a uniform transfer of heat to the retorts. The coke by-product from retorts was typically used to fuel producer ovens. Eventually clay or firebrick retorts replaced iron retorts, allowing much higher carbonization temperatures.

---

* Recent literature on manufactured gas plants has focused on the environmental burden remaining from decades of operation with improper storage or treatment of impurities such as tar.

FIGURE 5.3
Beds of retorts in a setting. (From Castaneda, C. J., *Invisible Fuel: Manufactured and Natural Gas in America, 1800–2000*, Twayne Publishers, New York, 1999.)

Gas was produced from beds of retorts in a sequence of batch reactions. Coal would be added through a door on one end of a retort, the door would be secured, and gas would be produced as the temperature was raised. As shown in Figure 5.2 and Figure 5.3, the gas produced in a retort would escape through ascension pipes, which ran outside of the setting up to a hydraulic main that was partially filled with circulating water. The ascension pipes ran at an angle above the hydraulic mains and then descended into the mains and below the surface of the water through "dip pipes." Gas from the dip pipes would bubble up through the liquid seal provided by the water, providing pressure regulation and removing some of the tar contained in the gas. Water from the hydraulic main ran off to a tar well, and the product gas was collected in a "foul main" that carried the gas to a series of washing processes and eventually to a gas holder. Figure 5.4 shows some of the main sequential processes, with a condenser, washer, and gas meter located between the retort bench and the gas holder. As shown, gas holders typically consisted of a large bell-shaped container inverted in water, with guide rollers to hold the bell in place as gas was added or removed. The weight of the bell maintained the stored gas at an increased and constant pressure. Each batch could take from 8 to 12 hours, depending upon the quantity of coal. The resulting coke product would be removed, typically being pushed out a door at the other end of the retort, and additional coal would be added for the next batch. This process was initially done by hand, as shown in Figure 5.5. Notice the rows of ascension pipes and the square producer inlets near the floor. The introduction of mechanical charging and discharging in large production plants significantly reduced labor costs.

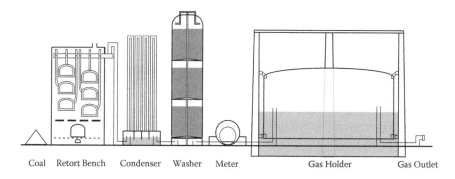

Coal   Retort Bench   Condenser   Washer   Meter   Gas Holder   Gas Outlet

FIGURE 5.4
Key elements of an early gasworks system. (Credit: M. Melaina, NREL.)

5.2.2.2 Water Gas and Carbureted Water Gas

A method of producing water gas was patented in France as early as 1834 and involved exposing steam directly onto incandescent carbon, usually coke or anthracite. The reactor was run in sequential steps, with one step heating the coke or anthracite by circulating producer gas and a subsequent step exposing steam to the incandescent coke or anthracite. The resulting fuel gas consisted of large fractions of hydrogen and carbon monoxide and had a heating value of about 300 Btu per cubic foot (see Table 5.1). According to a water gas

FIGURE 5.5
Workers loading retorts with coal. (From Stotz, L. and A. Jamison, *History of the Gas Industry*, Press of Stettiner Brothers, New York, 1938, 47.)

patent from 1867, water gas was composed of 47% hydrogen and 37% carbon monoxide (Castaneda 1999, p. 57). Water gas was relatively inexpensive to produce but burned with a blue flame that had little illuminating power, and therefore it initially had few markets (Tarr 1999, p. 22).

This limited applicability began to change after 1875, when Thaddeous S. C. Lowe patented a process of enriching water gas by adding gas produced by cracking liquid hydrocarbons, resulting in "carbureted" water gas, typically referred to simply as water gas. Carbureted water gas systems were more competitive than pure coal systems, requiring less space, capital, and labor, and having quicker start-up times and therefore reducing storage requirements. Carbureted water gas systems were often used to supplement coal gas systems rather than expanding a coal system by adding coal retort beds and storage capacity. The heating value of the mixed gas could be adjusted between 300 to 800 Btu/scf. The increased heating value of carbureted water gas resulted in increased candle power for the same quantity of gas at lower cost. Due to these advantages, carbureted water gas and mixed gas systems tended to displace pure coal gas systems.

5.2.2.3 Oil Gas

In addition to the increased use of oil products in the production of carbureted water gas through the Lowe process, oil also began to be used as the primary feedstock in gas production. Whale oil had actually been cracked to produce gas in the early 1800s, but in 1889 an improved refractory process was invented by L. P. Lowe, son of Thaddeous Lowe. The Pacific Coast was rich in oil and poor in coal, which proved to be a driver for the first oil gas plant constructed in Oakland, California, in 1902, and afterward this production process spread to other Pacific Coast cities (Tarr 2004, p. 736).

5.2.3 Urban Infrastructure Expansion

The schematic in Figure 5.6 shows major urban expansion patterns for manufactured gas and hydrogen infrastructure systems. The urban area on the left-hand side of the figure is served by an expanding manufactured gas system, and the urban area on the right-hand side is served by an expanding hydrogen infrastructure system. Three levels of expansion are indicated: (1) initial service, shown in black, (2) expanded service, shown in grey, and (3) introduction of long-distance transmission of natural gas and hydrogen, shown as dashed lines. Each city includes an urban center, shown as a shaded area. The manufactured gas system begins with an initial gasworks production plant and the delivery of gas through a pipeline network limited to the urban center. These pipelines would provide gas to streetlights to illuminate major roads and commercial areas, as well as to wealthy households. Urban centers in the 19th century included many wealthy households. Only after the introduction of mass transit and later automobiles did wealthy

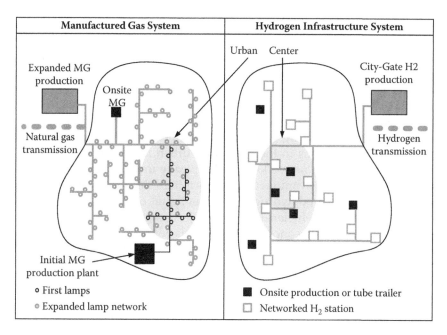

FIGURE 5.6
Comparison of city expansion patterns. (Credit: M. Melaina, NREL.)

households begin to shift away from city centers, which increasingly became dominated by businesses (Mohl 1985). On-site gas production facilities would have been installed to illuminate large buildings, such as factories or hotels, that were far removed from the initial pipeline system, as suggested by the black square in the northwestern section of the hypothetical manufactured gas city.

During the second level of expansion, a larger manufactured gas production facility is shown outside the city limits and provides gas to a larger pipeline network that both connects to and expands upon the existing urban core pipeline system. As the pipeline network expanded, it would have been more likely to reach remote buildings that had previously relied upon on-site production. This expanded network would involve higher volumes of gas and would achieve greater economies of scale in both distribution and production. The degree of connectivity between older and newer pipeline distribution systems would have depended upon the franchise agreements within particular municipalities. Unlike the system shown in this schematic, many franchises were relegated to well-defined territories and often competed for access to new customers and greater geographic coverage. As gas companies grew in size and influence, they began to become consolidated in urban areas across the country. Eventually, as indicated by the thick dashed line, long-distance pipelines would bring natural gas to the city gate where it would be distributed through the existing manufactured gas pipeline system.

A major distinction between manufactured gas and hydrogen infrastructures is the number of locations at which the fuel is required. For manufactured gas, service was provided on a household-by-household basis or on a streetlamp-by-streetlamp basis, thus requiring an extensive distribution system to survice many end-use points. This is indicated by the large number of circles in Figure 5.6 compared to the relatively sparse distribution of refueling stations in the case of an urban hydrogen infrastructure. Early hydrogen station networks would typically involve either on-site production systems (e.g., steam methane reforming [SMR] or electrolysis), stations receiving delivered hydrogen (e.g., via liquid or gaseous tank trucks), or mobile tube trailer refuelers (Ogden 1999). As suggested in the schematic, early stations will likely concentrate in urban centers, with some fraction of the total number of stations being dispersed across other parts of the city. In Figure 5.6, three on-site production or tube trailer stations are shown in the urban center and three are located in other areas during the first level of expansion. The second level of expansion is shown with a "city-gate" hydrogen production facility connected to a delivery truck or pipeline distribution network. This network connects multiple stations, including some of the initial on-site production or mobile refueling stations, which may or may not be converted to networked stations. Eventually, some large urban areas may be served by hydrogen transmitted from remote production facilities when large volumes of hydrogen are required. However, it has been suggested that significant quantities of various energy feedstock types could be economically delivered to the city gate and converted on-site (Myers, Ariff et al. 2003). This is only one scenario describing how an urban hydrogen infrastructure may evolve; many patterns of development are conceivable, and technological advances in the near term could very well influence which patterns are ultimately followed.

An intermediary phase missing from this schematic of an expanding hydrogen infrastructure is the potential transition from on-site production and truck delivery to large-scale production or long-distance pipeline transmission. Given that hydrogen refueling stations will dispense different volumes of fuel, grow at different rates, and steal market share from one another as the network of stations expands (Melaina and Bremson 2006), it is unclear how this transition might occur while efficiently allocating the capital needed for additional coverage and capacity. The manufactured gas industry was guided by franchises that addressed these types of intermediary expansion issues, partly by prescribing geographic territories within which companies were allowed to expand. In the case of hydrogen, addressing transition costs may prove more challenging due to the variability of demand at particular stations. Because manufactured gas pipes were plumbed directly to particular consumers, shifts in demand among end points were not a major concern. This analogy is simplified, but it highlights some of the important distinctions between manufactured gas and hydrogen infrastructure systems. The next section examines the major phases of manufactured gas infrastructure development in greater detail.

5.3 Phases of Infrastructure Development

The manufactured gas industry spanned roughly 130 years, beginning with the first gasworks company in Baltimore in 1816 and closing with the conversion to natural gas in the 1950s. The industry thrived during much of this history. But it was not static and adapted on several occasions in response to transformative internal and external technological innovations. The industrial world experienced many advances during the gas light era, including the first commercial railroad (1826), the telegraph (1836), the sewing machine (1844), the discovery of oil (1859), the telephone (1876), the incandescent light bulb (1879), the first airplane (1903), Henry Ford's assembly line (1913), and finally, and perhaps most importantly for the fate of the manufactured gas industry, the electric arc welding techniques that enable long-distance transmission of natural gas (1923).

The evolution of the manufactured gas industry can be discussed in terms of five major phases (c.f., Tarr 2004):

1. Demonstration to commercialization (1785–1816)
2. Early municipal adopters (1816–1850)
3. Coal gas dominance (1850–1882)
4. Expansion, competition, and adaptation (1882–1925)
5. Conversion to natural gas (1925–1954)

Before discussing these phases, a brief overview of the very early history of lighting technologies is warranted, if only to emphasize the significance of gas lighting as a major technological innovation. Before the industrial era, wax candles were the primary means of providing light. Tallow proved to be less expensive than wax, eventually replacing wax candles, and were themselves subsequently replaced to some degree by higher-quality whale oil lamps. Whale oil lamps provided the highest-quality lighting until the introduction of gas light. However, oil lamps were an age-old technology that had been used by primitive cultures with animal fat as a source of fuel. The superiority of whale oil lamps was primarily the result of discovering a cleaner fuel rather than an advance in lighting technology. As Nordhaus observes (1997), "There were virtually no new devices and scant improvements from the Babylonian age until the development of town gas in the late eighteenth century" (p. 34).

Oil lamps were relied upon for public street lighting for nearly 180 years before the introduction of gas streetlamps. Public lighting was provided in London in 1635 by lighting lanterns along city streets "before each tenth door" between sunset and midnight. The first public lighting in America was achieved in a similar manner in New York City, with a 1697 order by the city that every seventh resident "either place lamps in their front windows, or suspend one from a pole in front of the house." In 1761, the city of New York assumed responsibility for lighting oil streetlamps (Stotz and Jamison 1938, p. 114). It was against this history of lighting that the first demonstrations and commercial applications of gas lighting would develop in the late 1700s and early 1800s.

5.3.1 Demonstration to Commercialization (1785–1816)

The ability to conduct controlled demonstrations with gases was the result of scientific advances during the 18th century, but ad hoc demonstrations preceded these developments. Sir John Winter heated sea coal in 1656 in an earthenware pot, driving off sulfurous and arsenic gases (Smith 1945). As early as 1683, the English clergyman John Clayton had drained a spring that maintained a gas flame; digging under the site of the spring, Clayton found coal and subsequently experimented with the coal and captured flammable gas that he contained in ox bladders (Castaneda 1999). It had therefore become common knowledge that flammable gases could be produced from coal or wood by the time Henry Cavendish identified hydrogen in 1766, and before Alessandro Volta identified methane 10 years later. Lavoisier would challenge the phlogiston theory with his theory of oxygen combustion in 1777. Scientific theory, in this case, took more than a century to catch up with informal observations, and it would be decades more before controlled experiments lead to commercially viable applications. Some of the first "practical" applications of this new scientific knowledge of gases began with the ballooning craze, which began in 1783 when Jacques Alexandre César Charles made the first flight in a hydrogen-filled balloon in France. The manipulation of combustible gases eventually shifted from the laboratory, from the ballooning craze, and from ad hoc illumination demonstrations into the world of business and commerce. After commercial viability had been recognized, technical advances occurred more rapidly.

The first demonstrations and commercial applications of gas lighting occurred in Europe, though comparable American activities followed soon afterward. Credit as the "inventor of gas lighting" has been given to Jean Peirre Mincklers, who in 1785 illuminated his classroom at the University of Louvain with a gas flame (Elton 1958, cited in Castaneda 1999, p. 5). Around this time, William Murdock, a Scottish inventor, had been experimenting with distillation processes and had developed a gas-lighted headlamp for a steam-powered vehicle. In 1792, Murdock diverted coal gas through 70 feet of copper and tin pipes to light a room in his home in Redruth, Cornwall. Philippe Lebon had been experimenting during the same time period in France, and demonstrated his patented "thermolampe" in October 1801 by providing gas to light the Hotel Seignelay in Paris. This was perhaps the first high-profile gas light demonstration. Benjamin Henfrey demonstrated gas lighting technologies in Baltimore in 1802, employing a "thermo-lamp" design that was probably similar to that used by Lebon in France. In 1803 Henfrey installed a 40-foot tower with a gas light that illuminated part of a major street in Richmond, Virginia. In 1806, David Melville, an American pewterer and hardware merchant, installed a gas light system in his home, and in 1813 he attained a patent for a coal gas plant design. Melville installed several gas light systems in cotton mills between 1813 and 1817 in Massachusetts and Rohde Island. Both Henfrey and Melville recognized the practical application of gas lighting, but neither could attract adequate investments as a result of their efforts (Castaneda 1999, pp. 6–13).

The first business applications of gas lighting in Europe are often discussed in terms of the inventions, demonstrations, and business initiatives of four key players: William Murdock of Scotland and his English associate Samuel Clegg, Frederic Winsor of Germany, and Philippe Lebon of France. Lebon recognized the potential of gas lighting, but his efforts were cut short by his murder on the Champs Elysees in 1804, some three years after his demonstration in Paris. However, in response to encouragement from a business associate who had seen Lebon's demonstrations in France, Murdock redoubled his efforts and installed a gas light at each end of the Soho Works in 1802. A year later, the Soho Foundry was lit using a system of copper tubes and cockspur burners (Smith 1945). Not to be outdone by this building lighting demonstration, the world's first street lighting display was arranged by Frederic Winsor in London five years later in 1807, orchestrated to coincide with the King of England's birthday.

Winsor is generally credited with the idea of installing a large-scale gas light system that would cover multiple city blocks. Winsor had proposed to install pipelines under city streets in Germany and was refused, and he subsequently emigrated to England where he continued with his gas lighting ambitions. After the 1807 demonstration, Winsor sought an act of the English Parliament to tear up city streets in London for a gas light system. Murdock successfully opposed this application on the basis of right to priority and wanted to first establish gas lighting in industrial applications before introducing it in the residential sector. Winsor was eventually granted a joint-stock limited-liability charter on April 30, 1812, but the venture's progress was hampered, partly due to Winsor's lack of management and engineering skills, and he was eventually forced out of his own business by the board of directors. Under the leadership of Samuel Clegg, a former assistant to and later competitor with William Murdock, the same firm began installing pipe systems under city streets. Wooden pipes were used to deliver gas to Westminster Bridge, which was illuminated on New Year's Eve, 1813. By the end of 1815, the firm had established 26 miles of pipeline in London (Castaneda 1999, p. 8). Pipe began to be manufactured using the same methods used to produce gun barrels. By 1819, other English towns began to adopt gasworks systems, and London had nearly 300 miles of gas pipe supplying some 50,000 burners (Busby 1999, p. 6). Paris did not introduce gas lighting until 1820 (Speer 2008), and the first German gasworks began in Hanover in 1825 and Berlin in 1826 (Keller and Hoeferl 2007).

In contrast to the franchises that were granted in American cities, the charter granted to Winsor's Gas Light and Coke Company in 1812 did not allow for monopoly territory. There were four gas companies operating in London by 1823, and by 1857 there were three sets of mains under Cockspur Street and four companies supplying gas to Oxford Street. It was not until 1860 that Parliament sanctioned monopoly territories to eliminate excessive competition; this practice subsequently spread throughout England (Smith 1945, p. 13).

The gas light industry therefore first became firmly established in England, and much of the world, including major American cities, typically looked to England as a source of technical expertise. However, American gasworks developments followed quickly on the heels of English developments. The first substantial gas light businesses in the United States were established by the Peale brothers, Rembrandt and Rubens, whose father was a prominent portrait artist who had taken them to study painting in Europe in 1802, where they were exposed to gas street lighting demonstrations. The Peale brothers had a penchant for showmanship, and both established on-site gas lighting systems in museums. Rubens installed a coal-based gas lighting system in a closet under a stairway in his family's museum in Philadelphia in 1814. The system cost $600 ($9000 in 2007 dollars) and was modeled after Melville's design for cotton mills. This first system failed due to the noxious fumes that were inadvertently released into the interior of the museum. The second system, which used pine tar as a feedstock, cost $5000 ($75,000 in 2007 dollars) and was more successful (Castaneda 1999, p. 16). Eventually the museum was declared the first American building to be completely illuminated by gas light. A fee was charged to enter the museum and the gas lights proved to be a major attraction. Though the city council of Philadelphia was aware of Rubens's gas lighting systems, they offered no substantial support for additional projects. Rubens's brother Rembrandt Peale established a similar pitch-based gas light system in a museum in Baltimore in 1816, which had a similar effect of drawing visitors. More importantly, Rembrandt initiated the establishment of the first American gasworks company in 1816, the Gas Light Company of Baltimore. One of the first public buildings to be lit was the Belvidere Theater, located across the street from the new gasworks facility. The growth of gas lighting in Baltimore was gradual; by 1833 only 2 miles of pipeline were providing gas to 3000 private lamps for businesses and households and 100 public streetlamps (Castaneda 1999, p. 25). However, efforts in Baltimore paved the way for other cities to follow.

5.3.2 Early Municipal Adopters (1816–1850)

During this phase gasworks became established in major cities across the country and expanded rapidly within large cities such as New York and Boston. Municipal governments often determined the timing of introduction, which was limited by business interest and the negotiations leading to the granting of a franchise (as well as political resistance from whale oil and tallow interests, to some degree). Municipalities also set rates, illumination standards, and the duration of the franchise and often influenced the type of system installed. New York, for example, insisted on a range of detailed franchise conditions and required that the lead engineer visit gasworks in London (Stotz and Jamison 1938, p. 22). A gas light company started in Boston in 1822, and the New York Gas Light Company was established in 1823, though the actual introduction of gas occurred in each city years later (see Table 5.2).

With the viability of gasworks systems proven in these larger cities, which were themselves encouraged by successes in London and across England, other major cities followed in rapid succession. Introductions by major cities

TABLE 5.2

Municipal Early Adopters of Manufactured Gas

City	Year	City	Year
Baltimore, MD	1816	Columbus, OH; Hartford, CT; Worcester, MA	1850
New York City, NY	1825	Indianapolis, IN; Memphis, TN,	1852
Boston, MA	1829	Milwaukee, WI	1853
Louisville, KY	1832	San Francisco, CA; Toledo, OH	1854
New Orleans, LA	1835	Atlanta, GA	1856
Philadelphia, PA, Pittsburgh, PA	1836	Scranton, PA; St. Paul, MN	1857
Manchester, NH	1841	Portland, OR	1860
Cincinnati, OH	1843	Kansas City, MO; Los Angeles, CA; Oakland, CA	1867
St. Louis, MO	1846	Omaha, NE	1868
Fall River, MA; Newark, NJ	1847	Stockton, CA	1869
New Haven, CT; Paterson, NJ; Providence, RI; Rochester, NY; Washington, D.C.; Buffalo, NY	1848	Minneapolis, MN	1871
		Seattle, WA	1873
Chicago, IL; Norfolk, VA; Cleveland, OH; Detroit, MI; Syracuse, NY; Utica, NY	1849	Tacoma, WA	1885
		Spokane, WA	1887

Source: Compiled from Castaneda, C. J., *Invisible Fuel: Manufactured and Natural Gas in America, 1800–2000*, Twayne Publishers, New York, 27, 1999; Stotz, L. and A. Jamison, *History of the Gas Industry*, Press of Stettiner Brothers, New York, 9, 1938.

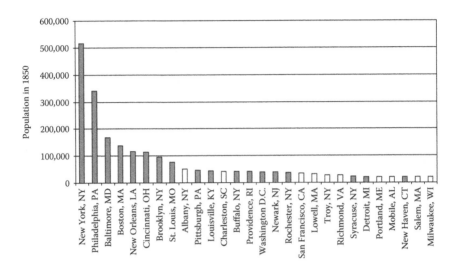

FIGURE 5.7

Large American cities ranked by populations in 1850. Cities with manufactured gas systems are highlighted. (Credit: M. Melaina, NREL.)

and by year are listed in Table 5.2. Castaneda (1999, p. 34) reports that some 50 plants had been installed by 1850, and a strong period of growth occurred after this year. However, this 1850 estimate is somewhat misleading because these 50 cities included the large majority of major American cities, so most urban residents had exposure to gas lighting before the strong growth in the total number of plants after 1850. Figure 5.7 shows the 28 largest American cities with populations greater than 20,000 persons in 1850, and highlights the 17 cities that had established manufactured gas systems (based upon cities listed in Table 5.2, which may be only a partial list). Because large cities tended to be early adopters, by 1850 approximately 78% of the 2.4 million persons living in American cities with populations greater than 15,000 persons had access or exposure to gas lighting.

5.3.3 Coal Gas Dominance (1850–1882)

Early gasworks systems relied upon a variety of feedstock types, but all large cities eventually settled on coal. The Baltimore Gas Light Company first used a pine tar system, the New York Gas Light Company first used whale oil and later rosin, and the Wilmington, North Carolina, and Philadelphia gasworks used wood (Tarr 2004, p. 734; Williamson vol. 1, p. 39). Clark reports that grease obtained from slaughterhouses was used in Dayton, Ohio, in 1848 (Clark, cited in Waples, p. 30). Coal was scarce and relatively expensive when these early systems were established, and coal carbonization resulted in significant impurities in the product gas. The purity issue was overcome with the importation of coal gas purification technology from England in the 1840s. Coal gas works on the East Coast used imported coal from England, while inland gasworks in Virginia, Kentucky, western Pennsylvania, and Illinois began using domestic coal (Williamson et al. vol. 1, p. 39). The price and scarcity issue was overcome with the completion of the railroad connection between Pittsburgh and Philadelphia in 1852. Given these developments, the diversity in production processes ended around 1850, after which coal carbonization became the dominant production method.

Other improvements during this period included the switch from cast iron retorts to refractory clay retorts, the use of dry meters rather than less reliable wet meters, and the manufacture of 12-foot sections of vertically cast iron pipe sections rather than 4- to 5-foot sections of horizontally cast iron pipe (Tarr 2004, pp. 734–735). As a result of these technical improvements, and the reduction in coal distribution costs, the number of reported gasworks increased rapidly, as indicated by the estimates in Figure 5.8. Beginning with around 50 plants in 1850, there were approximately 1500 plants by 1910. This strong growth occurred in spite of the panic and depression of 1857, and despite the American Civil War (1861–1865). As shown in Figure 5.8, the growth rate in the six decades after 1850 was much higher than during the first three decades of the industry after 1820. The downtown streetlamp projects financed by municipalities before 1850 were therefore just the first rumblings of what would later become a major industry.

Large cities such as New York and Chicago were served by multiple gas companies, each typically having one or more production facilities feeding

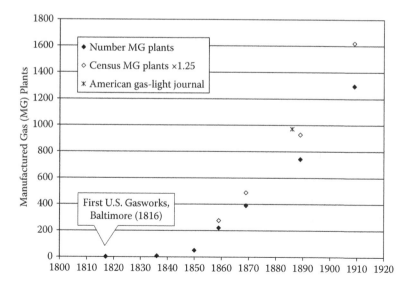

FIGURE 5.8
Growth in number of early manufactured gas plants.

a contractually and geographically defined distribution area. Small-scale gasworks were also established in many small towns (c.f., 5000 persons) for lighting, and small systems were installed at industrial facilities for either lighting or other processes (Tarr 2004, p. 734). Though the industry had spread to most major cities across the country by 1870, half of the total labor and capital in the gas lighting industry was located in four states: New York, Massachusetts, Pennsylvania, and Ohio (Tarr 2004, p. 735).

Gas companies had initially focused on public street lighting as their core business strength, but relatively stringent municipal franchises restricted profits from this market. Many firms either defaulted on their public lighting contracts or persistently petitioned city councils for rate increases for this service (King 1950, pp. 38–39, cited in Tarr 1999, p. 21). Gasworks ultimately attained most of their profits by supplying gas to businesses and private residences. However, the high visibility of public lighting served as a market building force for private consumers. It can be argued that investments in public lighting systems had significant indirect returns as a result of stimulating demand in private markets. Moreover, the technical improvements in distribution, production, burners and fixtures, and metering technology that occurred with the provision of public lighting systems resulted in reduced capital costs, higher efficiencies, and economies of scale that benefited companies as they expanded into private markets.

The industrial era would be fundamentally changed as a result of Colonel Drake's discovery of oil in Titusville, Pennsylvania, in 1859. The oil industry flourished between 1860 and 1910, primarily due to the sale of kerosene for lighting, which captured most of the lower- and mid-income household illumination markets during the 1870s and 1880s. Kerosene lamps were not of the same quality as gas lights, however, and therefore did not pose a direct challenge to the use of gas light by households who could afford higher-quality lighting. Moreover, with the advent of carbureted water gas, reductions in gas costs were achieved in new production plants, improving the competitiveness of manufactured gas. The 1890 census revealed that coal gas systems provided half of all manufactured gas, but this share dropped to nearly 10% by 1904, with the remainder being provided by carbureted water gas and mixed coal and water gas systems (Tarr 2004, p. 736).

Though kerosene did not prove a major threat, and oil-based carbureted water gas systems actually reduced the cost of gas lighting, real competition did materialize with Thomas Edison's Pearl Street power station and incandescent light bulb demonstration in 1882. At the same time, electric arc lamps also began to encroach on the street lighting market. But by 1880 the gas industry was receiving 90% of revenues from domestic lighting, and therefore arc lamps did not pose a significant threat (Tarr 2004, p. 736). Unlike arc lighting, the incandescent light bulb could be used indoors and provided a higher quality than the gas lighting systems of that time. In terms of the relative cost of the two lighting technologies, it has been suggested that American gas companies opted to maintain high prices and profits rather than reduce prices to compete directly with electric lighting, which allowed electric lighting to enter the market more quickly than in England, where, among other factors, lower gas prices made investments in electricity systems less appealing. The adoption of electric lighting in America appears to have been due to superior quality rather than relative costs (Shiman 1993). Regardless of which drivers were more fundamental, electric lighting grew rapidly and eventually dominated all urban lighting markets. During the same time period the oil industry would shift away from illumination markets and begin producing gasoline for vehicles. In response to pressures from these changing market conditions, the manufactured gas industry adapted and continued to expand through another period of strong growth.

5.3.4 Expansion, Competition, and Adaptation (1882–1925)

The technical superiority of electric lighting had been recognized by gas companies even before electricity was first introduced. Despite this recognition, and even as the threat from electric lighting became concrete, the gas industry continued to expand and innovate through a "golden era" of gas lighting. The challenge posed by electric lighting stimulated innovations in gas production and utilization and drove expansions into additional markets such as home heating and cooking, as well as industrial applications. Increased demands from these new markets maintained growth in the industry during the roaring 1920s, even as electric lighting captured greater market share. An Environmental Protection Agency study suggests that production capacity from large facilities increased by more than 10-fold between 1890 and 1930, from 24,000 MMscf to 365,000 MMscf (EPA 1985, p. 21). The three most significant technical innovations during this period were carbureted water gas, the Welsbach mantle, and improved storage technologies. At the end of this period the invention of arc welding for pipes would prove to be a transformative innovation, enabling the economical long-distance delivery of natural gas. Other important trends included the widespread adoption of residential gas heating and cooking, integration with the steel industry by way of the coke by-product oven, and the acquisition of electricity companies by gas companies.

5.3.4.1 *The Welsbach Mantle*

In 1885, Carl Auer von Welsbach invented the Welsbach mantle, an incandescent mantle that produced six times more light per quantity of gas than an open tip lamp. The Welsbach mantle was the gas lighting equivalent of the incandescent electric light bulb, which Thomas Edison had refined to endure for 1200 hours by 1880. The Welsbach mantle generated a bright white light by heating rare earth oxides such as thorium dioxide and cerium dioxide. In contrast, traditional gas lamps generated fire light based upon the emissivity of heated carbon particles. Generating light by heating a mantle rather than carbon particles made existing gas lamp candlepower standards obsolete; illumination was now a function of gas heat content rather than gas luminosity. As a result, beginning around 1908, and after intensive lobbying by the gas industry, municipalities began measuring gas light service in cost per energy, or BTU (British Thermal Unit), rather than volumes of gas at a specified Btu per standard cubic foot (Tarr 1999, p. 23). Incandescent gas street lamps grew in number during this period, beginning with 285 lamps in 1896 and growing to around 250,000 by 1914 (Stotz and Jamison 1938, p. 113). A coordinated, industry-wide effort to market incandescent lamps was largely successful, with 50 million incandescent gas mantles in use by 1914 (Stotz and Jamison 1938, p. 117).

5.3.4.2 Heating and Cooking

The use of gas for home cooking and heating had begun on a small scale by 1878, with about 12 gas stoves in operation in New York City and about 100 in operation across the country (Castaneda 1999). This was the same period of time in which incandescent electric lighting experiments were capturing attention from both the public and the business community. The gas industry became more unified and coordinated, forming the National Commercial Gas Association and accelerating market expansion through improved merchandizing methods (Stotz and Jamison 1938, p. 125). Advertising campaigns were launched to promote the industry, and retail outlets were established in prime downtown locations to demonstrate the latest advances in domestic cooking and heating appliances. The eventual result of these efforts was increased demand from household markets. The industry suffered from a lack of capital due to wartime restrictions during World War I but was then opened to public investors and experienced strong growth through the 1920s. Gas cooking was firmly established before the Great Depression, with use in "nearly 10 million homes" and sales of 1.4 million units in 1930 (Stotz and Jamison 1938, p. 170).

5.3.4.3 Integration with Steel

Industrial demand for gas as a fuel also contributed to growth of the industry during this period. Most industrial users could rely on lower heating value producer gas, which was typically produced on-site from coal (Steere 1922). Franchise agreements had required gas manufacturers to maintain heating value standards of around 450–550 Btu/scf to ensure adequate luminosity. This restriction was additional motivation to lobby for a change to measure gas in units of energy (i.e., Btus) rather than sales per volume with Btu/scf standards. This change in the standard allowed more gas to be sold to industrial users, and from 1919 to 1927 the use of manufactured gas in industrial applications approximately doubled, increasing from 70.4 billion to 136.4 billion scf (Tarr 1999, p. 25). During this same time period the steel industry was switching from the beehive coke oven to the by-product coke oven, which produced other valuable by-products in addition to gas and coke. The American industrial economy had developed to the point that there were established markets for these by-products, which included "tar used for creosote in railway tie preservation, pitch for footing and water-proofing, refined tar for road surfacing, and ammonia, cyanides or phenolic compounds for chemical industries" (Tarr 1999, p. 25). These markets had developed earlier in England, improving the economics of the first coal gas systems, but they had not existed during the introduction of coal gas in America. Contamination from coal tar spills at an estimated 3000–5000 former manufactured gas plants continues to be an environmental burden today. Some of these brownfield sites are owned by the modern successors of the original manufactured gas companies that built the plants. Interestingly, many of the site are located near downtown areas (EPA 2012).

During the 1920s, the gas industry became more integrated with the steel industry by focusing on high-quality coke production in by-product ovens. The marketing of multiple products allowed higher utilization of production facilities. By 1926 the steel and manufactured gas industries were relying upon very similar technologies. By 1932, companies operating coke production facilities were providing 25% of all manufactured gas for domestic, commercial, and industrial customers (Tarr 1999, pp. 25–26).

5.3.4.4 Merging with Electricity

Gas companies also responded to competition from electricity by purchasing electricity companies. The first acquisitions began as early as 1887, when approximately 40 manufactured gas companies were supplying electric lighting, mostly arc lighting for streetlamps. By 1899, about 40% of all gas companies were supplying both manufactured gas and electric lighting services. As large electricity companies became more extensive, they in turn began to purchase gas companies in the early 20th century (Tarr 2004, p. 738).

5.3.5 Conversion to Natural Gas (1925–1954)

Fredonia, New York, was the first American city to use natural gas for lighting in the 1820s, drawing upon gas from a natural spring (Castaneda 1999, p. 39). Large-scale natural gas discoveries coincided with the oil drilling frenzy that followed Drake's first oil well at Titusville in 1859, but the gas was typically vented and was considered a nuisance to drillers. Natural gas began to be used for industrial heating applications in the 1860s and 1870s and was used to distill brine water in Centerville, Pennsylvania, as early as 1840 (Castaneda 1999, p. 42). Pittsburgh, with a thriving iron industry, was the first city to use natural gas on a large scale; by the late 1880s there were 500 miles of natural gas pipeline serving residential and industrial customers in Pittsburgh, including 232 miles within the city and drawing from 107 regional gas wells.

Other cities located near gas resources followed Pittsburgh's example. Around 1900 there were four regions where natural gas was being imported into cities:

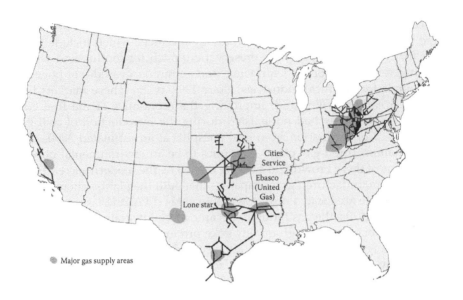

FIGURE 5.9
Major natural gas pipeline systems in 1929. (From Castaneda, C., and C. M. Smith, *Gas Pipelines and the Emergence of America's Regulatory State: A History of Panhandle Eastern Corporation, 1928–1993* [Studies in Economic History and Policy: USA in the Twentieth Century], Cambridge University Press, New York, 2003, 16.)

Pennsylvania and West Virginia, Northern and Central Indiana, locations around Los Angeles, and Eastern Kansas. Many cities that tapped into regional natural gas systems consumed the gas until wells ran dry, at which point they would revive their manufactured gas systems. As a result, the specter of uncertainty haunted natural gas for some time. With increased demand for petroleum, and uncertainty about the size and reliability of natural gas resources, coal gas systems experienced a temporary revival in the early 1900s.

Technological improvements in welding technology allowed for the long-distance transmission of natural gas beginning in the 1920s, and cities with existing natural gas systems expanded their pipelines to collect natural gas from more distant wells. The extent of natural gas pipelines serving these areas by 1929 is indicated in Figure 5.9. The uncertain reputation of natural gas changed with the development of large oil and natural gas fields in Texas and Oklahoma. A few major pipelines were installed before the Great Depression and the steel restrictions of World War II, which slowed growth in pipeline installations. Chicago began receiving natural gas in 1931 and Minneapolis in 1935. After World War II the expansion of natural gas transmission pipelines accelerated. The Big and Little Inch pipelines had been installed to transport petroleum to the Northeast during the war and were sold to natural gas companies in 1947, against strong protests by coal interests. By 1940, the natural gas transmission system had expanded considerably, as indicated in Figure 5.10.

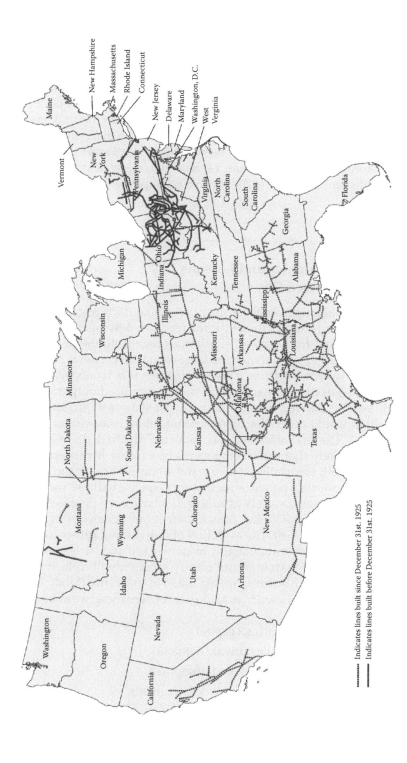

FIGURE 5.10
Extent of natural gas transmission pipelines in 1940. (From Castaneda, C. J., *Invisible Fuel: Manufactured and Natural Gas in America, 1800–2000*, New York, Twayne Publishers, 1999.)

Manufactured gas systems were initially supplemented by natural gas, and then eventually they were shut down as natural gas began to dominate urban markets. On a city-by-city basis, the urban distribution systems initially developed for manufactured gas were switched over to natural gas. Utilities launched large-scale conversion efforts to adjust household appliances to burn natural gas rather than manufactured gas. New York City was one of the last major cities to make the conversion and began receiving natural gas in 1951 (Castaneda 1999, p. 140). The remaining Northeastern markets followed soon afterward. Manufactured gas had been virtually displaced by 1965, and the large gas storage tanks, having been a symbol of large-scale urban energy systems, were dismantled in the 1970s and 1980s (Waples 2005, p. 40).

5.4 Conclusion: Analogies and Lessons from Manufactured Gas

There are several analogies from the history of the manufacture gas industry that offer lessons for future hydrogen infrastructure developments. Some of these lessons follow from general similarities regarding the networked nature of energy delivery systems and the large capital investments required to establish these systems. Economic theory suggests that monopoly conditions arise from (1) declining marginal costs with increased output and (2) system-wide benefits derived from common carrier status for infrastructure components that would otherwise be redundant under competition (Kahn 1990; Train 1991). The manufactured gas industry involved these characteristics, and it is likely that some components of a future hydrogen infrastructure will as well, especially pipelines and storage. There is significant debate over the appropriate policy or regulatory response to these and other monopoly conditions, and a variety of approaches have been proposed with the goal of ensuring that public and private benefits are achieved efficiently (Priest 1993; Williamson 1998). The evolution of current and future debates over the deregulation (or re-regulation) of networked industries, as well policy responses to climate change and energy security, will influence the manner in which hydrogen systems are regulated in the future (Mody 1996).

At least five lessons can be drawn from analogies between these two industries:

1. Long-term contracts can attract capital.
2. Public markets can stimulate private demand.
3. Competition breeds innovation and discipline.
4. Marketing evolves with technology developments.
5. Diversification can facilitate adaptation.

5.4.1 Long-Term Contracts Can Attract Capital

When private companies first began approaching municipalities requesting gas light franchise agreements in the 1820s and 1830s, many large municipalities had been funding public lighting via oil lamps for decades. The economics of whale oil lamps and gas lighting systems were fundamentally different, and gas lighting proved to be more amenable to the price stability, operational reliability, and illumination quality requirements sought by municipalities. Whale oil lamps incurred large operating costs and gas light systems incurred large up-front capital costs. The financial mechanism used to overcome these up-front capital costs was the issuance of long-term contracts, for as long as 30–40 years. The combination of contractual rates, monopoly territories, growing demand, and declining marginal costs made gas light companies attractive investments, and private investors gained significant returns in both Europe and America. Franchise agreements often evolved into negotiable contracts with some degree of flexibility, especially as market territories expanded and new market entrants were able to offer less expensive gas based upon improved technologies. Negotiations over issues such as rates, luminosity, heating values, and service territories resulted in heated local political battles (Tarr 2004, p. 734). In the 1880s, when competition from electric lighting was imminent, many gas light contracts established during the post-1850 growth phase were reaching the end of their terms.

Long-term contracts would obviously improve the economic standing of hydrogen systems as well, though such contracts are not likely to arise without the realization of firm future demand. This demand was established with municipal street lamps, which had been maintained and funded by some municipal governments for a century or more before the introduction of gas lighting. In the case of fuel cell vehicles, the uncertainties of future hydrogen vehicle cost reductions (e.g., fuel cell and hydrogen storage costs) and the uncertainty of the relative advantages of competing advanced vehicle fuels and powertrains (e.g., electric hybrids or advanced biofuels) make long-term contracts unlikely without strong government support for either vehicles or hydrogen (c.f., Greene, Leiby et al. 2008). Fleets of fuel cell transit buses may resemble the street lamp analogy, and some researchers have proposed central fleets as a stepping-stone to larger household vehicle markets (Kemp, Schot et al. 1998; cf. Nesbitt and Sperling 1998). But the extent to which niche household markets are willing to pay a premium for hydrogen vehicles, as did wealthy households for gas lamps, may prove to be the most relevant factor for financing major expansions in hydrogen fueling infrastructure. As early fuel cell and other electric vehicles are deployed, additional market research will reveal the degree and extent to which private households may place a premium on sustainable electric-drive vehicles.

5.4.2 Public Markets Can Stimulate Private Demand

Early municipal franchise agreements focused on supplying illumination for city streets. These early streetlight contracts were favorable enough to attract private capital, but streetlights also played a key role in the development of private household markets. Individuals—and in particular wealthy individuals—frequenting downtown streets and business districts were exposed to the benefits of gas lighting on a regular basis. With time, household markets proved much more lucrative than public streetlight contracts, and gas light companies sustained their growth primarily through household lighting profits. Despite these profits, the industry continued to serve primarily wealthy households; lamps fueled by whale oil, coal oil, and later kerosene were cheap alternatives to gas lighting, and therefore dominated lower-income markets.

Analogies for hydrogen in this regard can be drawn with the use of fuel cell vehicle technologies in either controlled fleet applications that are publicly funded or in private fleets with high fuel costs where investments in more efficient vehicles are economically attractive. Hydrogen and fuel cell buses have already been deployed in demonstration transit bus and shuttle bus fleets, where high ridership rates result in a large number of persons being exposed to hydrogen vehicle technologies. A similar awareness-raising influence could be achieved in light- and medium-duty fleets such as taxis or delivery trucks, which operate in high-visibility public spaces. The analogy can be taken further given the learning-by-doing cost reductions that can be attained through demonstrations and deployment of limited-production vehicle platforms. Network effects could reduce costs for stations serving both controlled fleets and private vehicles, such as the recent hydrogen station located at a transit bus depot in Emeryville, California (CaFCP 2012). Similar learning-by-doing and network cost reductions also resulted from early streetlight applications, enabling manufactured gas companies to offer lower rates to private households.

5.4.3 Competition Breeds Innovation and Discipline

Between the establishment of multi-decade municipal franchise contracts in the mid-1800s and Edison's demonstration of electric lighting in 1882, the manufactured gas industry enjoyed monopoly conditions and no major challengers to the high end of the household lighting market. And even after the gas companies had realized the "writing on the wall" with regard to the technological advantages of electric lighting, the industry continued to experience a golden age of growth into the early 1900s. These conditions came to an end as electric lighting gained market share, and the gas industry ultimately shifted away from illumination markets and into home heating and cooking markets and industrial markets.[*] Before this shift, the indus-

---

[*] For an intriguing discussion of the competitive interaction of gas and electric lighting, see Shiman 1993.

try responded with technological innovations and institutional focus. The Welsbach mantle (incandescent gas mantle) improved the quality of gas lighting and reduced costs, gas storage systems improved, and advances in coke ovens allowed gas to be a by-product from the production of coke for the steel industry. Institutionally, the gas industry aligned with trade organizations that helped to build new markets for gas, especially through public outreach promoting gas-powered household appliances such as cooking ranges, water heaters, and even refrigerators. Gas companies moved into the electricity business. Technologically, carbonization reactors (i.e., coke ovens) became the province of the steel industry.

An analogous competitive influence for hydrogen vehicles can be seen in the challenge posed by advanced vehicle power trains such as gasoline hybrid-electric and plug-in vehicles, and by advanced biofuels such as cellulosic ethanol or drop-in biofuels. These vehicle and fuel technologies can offer many of the same (public) advantages offered by hydrogen vehicles, including displacement of petroleum use, reduced criteria emissions, and reduced life-cycle greenhouse gas emissions. Interestingly, these vehicle technology developments tend to shift the burden of reducing impacts onto fuel providers rather than vehicle manufacturers. For example, hydrogen, electric, or biofuel vehicles can still have significant greenhouse gas emissions, depending upon how the fuels are produced. The "competition" to provide low-impact or sustainable personal vehicle travel is therefore multifaceted. Moreover, there is significant potential for overlapping influences, such as the shared electric drive train components between fuel cell and plug-in hybrid vehicles, the potential to develop internal combustion hydrogen or plug-in fuel cell vehicles, or the direct and indirect land use impacts of fuels produced from biomass resources. As was the case with manufactured gas and electricity, the relative public and private advantages of these different fuels and vehicle platforms will influence their market viability. Institutional strategies and technological synergies may also prove influential.

5.4.4 Marketing Evolves with Technology Developments

Early gas lamp households were billed by the number of lamps installed in each home. Gas inspectors roamed city streets announcing "lights out" and would shut down inlets if lamps were not extinguished after certain hours. Illumination was measured in candle-hours, and franchise contracts typically stipulated acceptable ranges for this metric, which correlated with the heat content of gas for open tip lamps using gas produced from coal carbonization. During this early period, gas companies owned residential light fixtures, which were typically imported and therefore expensive. After gas companies began manufacturing their own fixtures, it took some effort to convince homeowners to purchase fixtures, and initially they were financed. Later, after improvements in gas metering technologies, gas was sold on a cubic foot basis, a development which most certainly improved conservation since costs had previously only been roughly associated with actual gas usage (Stotz and Jamison 1938, p. 35).

The Welsbach mantle disrupted 80 years of industry practice by making candle-hour requirements obsolete. Incandescent mantles produced light by converting heat rather than by relying upon the luminosity of the fuel, which allowed lower heating value gas to be used for lighting, such as gas mixtures including carbureted water gas. Gas also began to be used for residential heating and cooking, and as an industrial fuel. In 1908 these trends culminated in the first heating value standard, or Btu standard, with New York City being the last municipality to convert to the new standard in 1922 (Stotz and Jamison 1938, 141). Eventually the industry adopted the therm, equal to 100,000 Btus, as the standard unit of measurement.

The manufactured gas industry had to adjust marketing methods and metrics in response to evolving technologies and expansion into new markets. Similar adjustments may be needed over time in the marketing of hydrogen for vehicles. Currently, the cost of hydrogen for vehicles is typically discussed in terms of dollars per kg of hydrogen, on the rationale that 1 kg of hydrogen has approximately the same energy content as a gallon of gasoline. However, because fuel cell vehicles use hydrogen more efficiently than vehicles normally use gasoline, this basis is somewhat misleading; if a fuel cell vehicle drives twice as far on a kg of hydrogen than does a gasoline vehicle on a gallon of gasoline, then a consumer could pay twice as much for hydrogen on an energy basis and still be paying roughly the same on a per-mile-driven basis. Moreover, the fuel cell vehicle owner has (probably) already paid a higher price up front for a more efficient vehicle, which is further justification for a fuel price signal that approximates the actual per-mile cost. Moreover, some early fuel cell vehicle owners will likely pay a premium on "green" hydrogen, adding another layer of complexity to the marketing of hydrogen. As vehicle and hydrogen production technologies change over time, and as policies promoting certain types of fuels are implemented (e.g., low carbon or sustainable fuel standards), both fuel providers and vehicle manufacturers may need to adjust how their products are marketed.

5.4.5 Diversification Can Facilitate Adaptation

The manufactured gas industry began with a diversity of production processes, including wood, rosin, and whale oil systems, but eventually became dominated by coal gas systems by around 1850. Additional production diversity was introduced with the advent of water gas. The usefulness of this production process was reinforced by the carbureted water gas system, which mixed water gas with light hydrocarbons created by thermally cracking different oil products. Pure oil production processes were also introduced and tended to dominate in West Coast cities where oil resources were plentiful. Another aspect of production diversity was introduced with increased use of natural gas, which began as a local or regional supplement to manufactured gas systems and later dominated the entire gas industry when larger fields were discovered and long-distance pipelines were established. Mixed production systems, such as combined coal and carbureted water gas systems, could in theory switch between feedstock types in response to market signals. This was seen in some cities where coal plants were revived after regional natural gas supplies ran dry.

End-use diversity was also introduced over time as the manufactured gas industry increased efforts to market to industrial and residential consumers. This marketing effort occurred largely in response to the threat posed by electric lighting and was enabled by the switch from a luminosity-based standard to an energy-based standard for gas contracts. The use of gas in multiple applications tended to balance out daily demand loads, allowing existing facilities to be used at higher utilization rates. The shift toward coke by-product ovens to serve industrial applications also resulted in increased revenues from refined coke for steel and from by-products such as tar compounds and ammonia. Diversity in both production and end-use applications helped to ensure continuity as the gas industry shifted away from the illumination business.

An analogy can be drawn here to the diverse methods that can be used to produce hydrogen. Conversion of natural gas, via steam methane reforming, and electrolysis are the two main methods of producing hydrogen today, but virtually any primary energy resource can conceivably be converted into hydrogen, as is the case for electricity. The three major low-carbon methods of producing hydrogen are from fossil fuel resources with carbon sequestration, from nuclear energy, and from renewables (Ogden 1999; Turner, Sverdrup et al. 2008). Therefore, even in a carbon-constrained future, hydrogen can maintain a significant degree of diversity in production methods. Potential applications for both stationary and mobile fuel cell applications add an additional degree of diversity to future hydrogen fuel cycles.

References

Bouman, M. J. (1987). Luxury and Control: The Urbanity of Street Lighting in Nineteenth Century Cities. *Journal of Urban History* 14(November): 7–37.

Busby, R. L. (1999). *Natural Gas in Nontechnical Language*. Tulsa, OK, PennWell.

CaFCP. (2012). Sneak Peek: AC Transit Emeryville Station. California Fuel Cell Partnership Retrieved February 17, 2012, from http://www.brownfieldstsc.org/roadmap/spotlight12.cfm.

Castaneda, C. J. (1999). *Invisible Fuel: Manufactured and Natural Gas in America, 1800–2000*. New York, Twayne Publishers.

Castaneda, C. J. and C. M. Smith. (2003). *Gas Pipelines and the Emergence of America's Regulatory State: A History of Panhandle Eastern Corporation, 1928–1993* (Studies in Economic History and Policy: USA in the Twentieth Century). Cambridge University Press, p. 16.

Clark, J. A. (1963). *The Chronological History of the Petroleum and Natural Gas Industries.* Houston, TX, Clark Book Co.

Elton, A. (1958). *Gas for Light and Heat. A History of Technology: The Industrial Revolution.* C. Singer. New York, Oxford University Press: 258–259.

EPA. (1985). *Survey of Town Gas and By-Product Production and Locations in the U.S. (1880–1950).* U.S. Environmental Protection Agency, Office of Environmental Engineering and Technology, EPA-600/7-85-004.

EPA. (2012). Remediating Manufactured Gas Plant Sites: Emerging Remediation Technologies. U.S. Environmental Protection Agency. Retrieved February 16, 2012, from http://www.brownfieldstsc.org/roadmap/spotlight12.cfm.

Greene, D. L., P. N. Leiby, et al. (2008). *Analysis of the Transition to Hydrogen Fuel Cell Vehicles & the Potential Hydrogen Energy Infrastructure Requirements.* Oak Ridge National Laboratory, from http://www-cta.ornl.gov/cta/Publications/Reports/ORNL_TM_2008_30.pdf.

Kahn, A. E. (1990). *The Economics of Regulation: Principles and Institutions.* Cambridge, MA, MIT Press.

Keller, F. and A. Hoeferl (2007). *Fighting for Public Services: Better Lives, a Better World,* Public Services International.

Kemp, R., J. Schot, et al. (1998). Regime Shifts to Sustainability Through Processes of Niche Formation: The Approach of Strategic Niche Management. *Technology Analysis & Strategic Management* 10(2): 175–195.

Melaina, M. W. and J. Bremson (2006). *Regularities in Early Hydrogen Station Size Distributions. Energy in a World of Changing Costs and Technologies,* 26th North American Conference, International Association of Energy Economics, Ann Arbor, Michigan.

Mody, A., Ed. (1996). *Infrastructure Delivery: Private Initiative and the Public Good.* Washington, DC, The World Bank.

Mohl, R. A. (1985). *The New City: Urban American in the Industrial Age, 1860–1920.* Arlington Heights, IL, Harlan Davidson, Inc.

Myers, D. B., G. D. Ariff, et al. (2003). *Hydrogen from Renewable Energy Sources: Pathway to 10 Quads For Transportation Uses in 2030 to 2050.* Directed Technologies, Inc., Arlington, Virginia. Grant No. DE-FG01-99EE35099.

Nesbitt, K. and D. Sperling (1998). Myths regarding Alternative Fuel Vehicle Demand by Light-Duty Vehicle Fleets. *Transportation Research Part D: Transport and Environment* 3(4): 259-269.

Nordhaus, W. D. (1997). Do real-output and real-wage measures capture reality? The history of lighting suggests not. *The Economics of New Goods: History, Theory, Methodology and Applications.* T. F. Bresnahan and R. J. Gordon. 1, University of Chicago Press.

Nye, D. E. (1999). *Consuming Power: A Social History of American Energies.* Cambridge, MA, MIT Press.

Ogden, J. M. (1999). Prospects for Building a Hydrogen Energy Infrastructure. *Annual Review of Energy and the Environment* 24: 227–279.

Priest, G. L. (1993). The Origins of Utility Regulation and the "Theories of Regulation" Debate. *The Journal of Law & Economics* 36(1): 289–323.

Schivelbusch, W. (1988). *Disenchanted Night: The Industrialization of Lights in the Nineteenth Century*. Berkeley, University of California Press.

Shiman, D. R. (1993). Explaining the Collapse of the British Electrical Supply Industry in the 1880s: Gas versus Electric Lighting Prices. *Business and Economic History* 22(1): 318–327.

Smith, N. (1945). *Gas Manufacture and Utilization*. London, The British Gas Council.

Speer, D. (2008). The Lamplighter of Olde Middle Village. Retrieved October 5, 2008, from http://www.junipercivic.com/HistoryArticle.asp?nid=13.

Steere, F. W. (1922). Producer Gas Technology. *American Fuels*, Vol 2. R. F. Bacon and W. A. Hamor. New York, McGraw-Hill.

Stotz, L. and A. Jamison (1938). *History of the Gas Industry*. New York, Press of Stettiner Brothers.

Tarr, J. A. (1999). Transforming an Energy System: The Evolution of the Manufactured Gas Industry and the Transition to Natural Gas in the United States (1807–1954). *The Governance of Large Technical Systems*. O. Coutard. New York, Routledge: 19–37.

Tarr, J. A. (2004). History of Manufactured Gas. *Encyclopedia of Energy* 3: 733–743.

Train, K. E. (1991). *Optimal Regulation: The Economic Theory of Natural Monopoly*. Cambridge, MA, The MIT Press.

Turner, J., G. Sverdrup, et al. (2008). Renewable Hydrogen Production. *International Journal of Energy Research* 32: 379–407.

Waples, D. A. (2005). *The Natural Gas Industry in Appalachia: A History from the First Discovery to the Maturity of the Industry*. Jefferson, NC, McFarland & Company, Inc.

Williamson, O. E. (1998). The Institutions of Governance. *The American Economic Review* 88(2): 75–79.

6

Fuel Cell Technology Demonstrations and Data Analysis

Jennifer Kurtz, Keith Wipke, Leslie Eudy, Sam Sprik, and Todd Ramsden

CONTENTS

6.1 Introduction

U.S. government–funded hydrogen and fuel cell demonstrations support technology research and development, and researchers at the National Renewable Energy Laboratory (NREL) are working to validate hydrogen and fuel cell systems in real-world settings. A key component of these demonstrations and deployments involves data collection, analysis, and reporting. NREL's Hydrogen Secure Data Center (HSDC) was established in 2004 as the central location for data analysis and works with DOE and its fuel cell award teams to collect and analyze data from these early deployment and demonstration projects. We strive to provide an independent third-party technology assessment that focuses on fuel cell system and hydrogen infrastructure performance, operation, maintenance, and safety. The analysis is regularly updated and published by application and is summarized in this chapter.

The U.S. Department of Energy (DOE) funds many demonstration and deployment projects in multiple application categories. DOE has designated more than $40 million from the American Recovery and Reinvestment Act (ARRA) of 2009 for the rapid deployment of approximately 1000 fuel cell systems across the United States. These are intended to facilitate the development of fuel cell technologies, manufacturing, and operations in strategic markets—material handling equipment (MHE), backup power (BU), and stationary power—where fuel cells can compete with conventional technologies.

The U.S. Department of Defense (DOD) is also funding fuel cell demonstration projects for MHE and BU. DOD's Defense Logistics Agency (DLA) is conducting a multiyear pilot project to demonstrate hydrogen fuel cell–powered forklifts for its MHE operations in warehousing and distribution centers. This pilot project encourages the deployment of fuel cells at volumes that can lower costs, promote market acceptance, and produce data to validate technologies in real-world applications.

DOE began the National Fuel Cell Electric Vehicle (FCEV) Learning Demonstration in 2004 to conduct an integrated field validation that simultaneously examines the performance of fuel cell vehicles and hydrogen infrastructure. More than 152 FCEVs have been deployed, and 24 fueling stations were placed in use.

Since 2000, NREL has evaluated fuel cell bus (FCB) demonstrations, including buses, infrastructure, and implementation experience. These evaluations were funded by DOE and the U.S. Department of Transportation's (DOT) Federal Transit Administration (FTA).

6.2 Government-Funded Fuel Cell Application Demonstrations and Deployments

The HSDC has evaluated fuel cell systems in light-duty automotive, heavy-duty transit, hydrogen production and dispensing, material handling, and backup power. Other applications that will be evaluated (but not discussed

TABLE 6.1

Summary of FC Application Benefits and Key Performance Areas

Application	Benefits	Key Performance Areas
FCEV	Zero-emission vehicle with renewable hydrogen production Increased efficiency Demonstrated range of 250 miles Decreased dependence on foreign oil Capable of meeting market demands	Durability, driving range, cost
FCB	Zero-emission vehicle with renewable hydrogen production Increased efficiency Noise reduction Decreased dependence on foreign oil Capable of meeting market demands	Durability, reliability, cost, weight reduction
FCMHE	Productivity gains such as decreased downtime Ease of use and maintenance Low or zero emission Steady power Capable of meeting market demands	Durability, efficiency, reliability, cost
FCBU	Reliability Low-noise Low or zero emissions Remote monitoring Capable of meeting market demands	Reliability, cost

here) are stationary prime power, portable power, and auxiliary power. The applications and hydrogen fuel cell systems discussed are from the U.S. systems that have been (or are being) evaluated at NREL's HSDC. Analysis results are published at least twice per year and can be found at the NREL website (6). (See Table 6.1 for a summary of benefits and key performance areas.)

More than 470 fuel cell systems (Figure 6.1) and 40 hydrogen production or dispensing locations have been analyzed to date; approximately 800 more systems will come on line in the next two years. The fuel cell systems are operating in many U.S. locations (Figure 6.2).

More than 410,000 hours have been accumulated, 3.3 million miles traveled, 66,000 hydrogen tank fills, and 190,000 kg of hydrogen produced or dispensed, or both (see Figure 6.3). The FCEVs and FCBs have been operating a few years longer than the FCMHE units, but the FCMHE units have already accumulated more than 251,000 hours with less than 10% of the total hydrogen. The FCBs have the fewest systems but still comprise a significant piece of the accumulated hours, hydrogen amount, and miles traveled, because of the lengthy driving profiles in transit revenue service. The FCBU units are

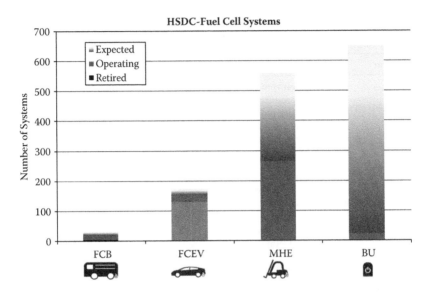

FIGURE 6.1
Summary of fuel cell applications and count analyzed in HSDC.

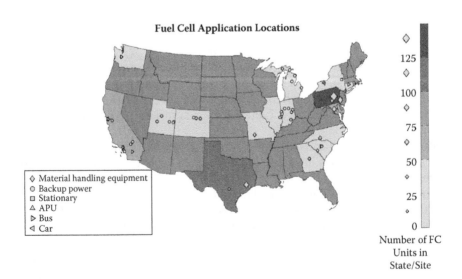

FIGURE 6.2
Fuel cell quantities and locations by application.

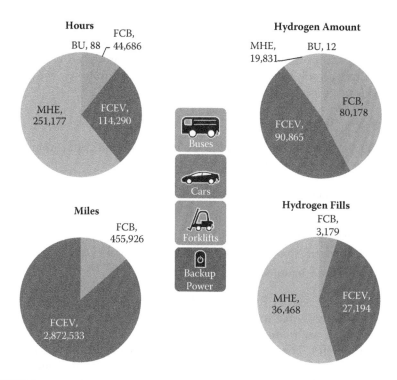

FIGURE 6.3
Summary of H2 FC hours, miles, and hydrogen amounts analyzed in HSDC.

a small percentage of the combined values, because they do not regularly operate for long periods.

6.2.1 Light-Duty Automotive

FCEVs offer a number of benefits in the light-duty automotive application, specifically zero-emission vehicles that do the following:

- Use hydrogen produced from renewable sources
- Have a demonstrated range of at least 250 miles
- Have high efficiency
- Provide the flexibility, size, and range the market demands
- Decrease our dependence on foreign fuels

A transition occurred with the makeup of the industry participants who initiated the FCEV Learning Demonstration in 2004. This is reflected in the number of retired vehicles (Figure 6.1) and stations. The projects started with four automotive original equipment manufacturer (OEM) and energy partner teams. Two project teams provided their last data by early 2010. We have analyzed data from more than 436,000 individual vehicle trips covering 2,800,000 miles and more than 90,000 kg hydrogen produced or dispensed, or both. Our key objectives are to evaluate fuel cell durability, vehicle driving range, and on-site hydrogen production cost.

6.2.2 Heavy-Duty Transit

FCBs offer a number of benefits in the heavy-duty transit application, specifically low- or zero-emission buses that do the following:

- Use hydrogen produced from renewable sources
- Have high efficiency
- Have decreased noise
- Meet the performance requirements of revenue service
- Decrease our dependence on foreign fuels.

NREL first evaluated FCBs in 2000 for SunLine Transit Agency in the Palm Springs, California, area and now has six active FCB projects. Additional FCB evaluations are planned for DOE and the FTA National Fuel Cell Bus Program (NFCBP). The NFCBP initiated in 2007 a $49-million, multiyear, cost-share research program to develop and demonstrate commercially viable fuel cell technologies for transit buses. The program included FCB demonstrations, component development projects, and outreach projects. The FTA is expanding the NFCBP with an additional $13.5 million in Bus and Bus Facilities funding from the FY 2010 DOT Appropriations Bill. FTA has solicited project proposals for these funds and additional funds that may become available. So far more than 455,000 miles have been traveled and 80,000 kg of hydrogen have been produced or dispensed, or both.

6.2.3 Material Handling Equipment

FCMHEs offer a number of benefits in the material handling application, specifically in productivity gains that can meet market demands, such as the following:

- Decreased downtime for hydrogen fills
- Steady power delivered
- Ease of use and maintenance
- Low or zero emissions

Since 2009 NREL has analyzed FCMHEs operating at real-world sites to understand their operation and performances. Although it is still too early to identify and understand operation trends, the number of operating

FCMHEs has increased from a few to more than 260 in the last 18 months because of DOE and DoD funding. An estimated 290 units are expected to be in operation in the next year. Data collected will provide great insight into the technology status and progress. Most operating units have already accumulated more than 1000 hours. We have analyzed more than 197,000 data files from FCMHE units with more than 251,000 hours and 19,800 kg of hydrogen dispensed.

6.2.4 Hydrogen Fuel Dispensing

We have analyzed many types of hydrogen infrastructure sites that include delivered and on-site production (natural gas reforming and electrolysis) capabilities. We have reported more than 190,000 kg of hydrogen produced or dispensed, or both, with more than 66,000 fills and only five infrastructure safety incidents.

6.2.5 Backup Power

FCBUs offer a number of advantages to the backup power application that can meet market demands, specifically the following:

- Reliability
- Low or zero emissions
- Ease of use and maintenance
- Remote monitoring

Since late 2009 NREL has analyzed FCBUs operating at real-world sites. FCBUs operate infrequently but are expected to have a very high reliability. So far, the 24 units we have analyzed have accumulated 88 run hours and had 199 successful starts out of 201 attempts (99% reliable). It is too early to accurately evaluate operation trends, but the number of deployed FCBU systems will soon increase to several hundred.

6.3 Approach

Some applications have specific objectives; one of our general objectives is to independently assess the technology status and progress for hydrogen and fuel cell components. We can accomplish this by structuring our analysis around highly collaborative relationships with our industry partners and stakeholders. The project partners and/or sites send us data every three months that enable us to perform unique and valuable analyses across all

FIGURE 6.4
HSDC data flow diagram.

the data in an application and across applications. We established the HSDC to protect the commercial value of these data for each company (Figure 6.4).

To ensure value is provided to the hydrogen, fuel cell, and end user communities, we publish *composite data products* (CDPs) twice per year at technical conferences to report on the progress of the technology and the project, focusing on the most significant and recent results. These CDPs will help these communities thoroughly understand the state of fuel cell technologies without revealing proprietary data. As such, they will provide useful feedback about technology status and fuel cell performance.

We also provide nonpublic detailed analyses of data for each company to maximize the benefit to industry of NREL's analysis work and to obtain feedback about our methodologies and results. These are called *detailed data products* (DDPs). Operation data collected at each deployment site includes performance, maintenance, fuel, and cost data. The publicly available CDPs are disseminated through the NREL website, conference presentations, and papers. In general, our data analysis objectives for all applications are to do the following:

- Independently assess technology by focusing on fuel cell system and hydrogen infrastructure performance.
- Support the growth of fuel cell technologies by reporting on technology features relevant to the business case and establishing a record of real-world fuel cell operation and maintenance data.
- Report on technology status to fuel cell and hydrogen communities, government, and end users.
- Identify barriers and areas for improvement.
- Leverage data processing and analysis capabilities from the fuel cell applications.

These deployments, combined with the data analysis, will facilitate the development of fuel cell technologies, manufacturing, and operation in strategic

FIGURE 6.5
Introductory screen of NREL's Fleet Analysis Toolkit.

markets where fuel cells can compete with conventional technologies. An increase in the number of fuel cell systems purchased will pave the way toward production-scale fuel cell manufacturing, which will in turn reduce system costs.

To evaluate large datasets, NREL developed an in-house tool called the Fleet Analysis Toolkit (NRELFAT), which helped researchers organize and automate the analyses performed on vehicles, systems, and infrastructure. Figure 6.5 shows the initial NRELFAT screen. The tool has recently undergone a major rework to extend the analysis functions to FCBs, FCMHEs, FCBUs, laboratory fuel cell data, stationary fuel cells, and plug-in hybrid vehicles. Having such a sophisticated tool in-house allowed us to rapidly respond to DOE's and DOD's needs for evaluation of early market fuel cell applications.

6.4 Results

Many of our analysis results cover a number of data metrics, not all of which will be included here. All published data, presentations, and papers can be found online (6). Some of the following data topics will be for an individual application; others will cover multiple applications. (Refer to the icon in the lower left corner of each figure for an indicator of the application data source.) So far 80 CDPs have been published for the FCEVs, 26 for the FCMHEs, and 6

TABLE 6.2

Analysis Topics Covered in Report by Application

Topic	FCEV	FCB	FCMHE	FCBU
Operation Hours	X	X	X	
Durability	X			NA
Hydrogen Fill Rate	X	X	X	NA
Fuel Economy	X	X	NA	NA
Range	X		X	NA
Hydrogen Produced/Consumed	X		X	
Efficiency	X			
Reliability		X		X

for the FCBUs. Additional CDPs are being conceived as additional trends and results of interest are identified, and as we receive requests from DOE, industry, and codes and standards committees. For the FCMHEs and FCBUs that are still in the early stages of deployment and data collection, identifying trends is difficult. We do expect results to be published in the coming year that will address performance metrics that are important to the value proposition of FCMHEs and FCBUs. See Table 6.2 for a list of the results summarized in this report, by application. Not all applications are covered in the results subsections because the analysis topic is not applicable to the application or data results were not yet available for the application. Data represents operation through July 2010.

6.4.1 Operation Hours (Combined FCB, FCEV, FCMHE)

The analyzed FCEVs started deployment in 2005, and fewer than half of the few fuel cell systems have accumulated more than 500 hours. Some have accumulated thousands of operation hours, but it takes time to accumulate so many hours, as more than 47% of the FCEV trips travel 20 or fewer miles per day. FCEV hours compared with FCMHEs (which started deployment in 2009) hours can be seen in Figure 6.6. Most FCMHEs have already surpassed 1000 hours. Operation of FCMHEs differs from FCEVs because FCMHEs may operate many hours a day for up to seven days a week, depending on the site.

6.4.2 Durability (FCEV)

Many improvements have been made in NREL's fuel cell durability analysis methodology, including using a two-segment linear fit and a weighting algorithm to calculate a more robust and automatic fleet average. Now that the data submissions are complete on first-generation (Gen 1) stacks (no new Gen 1 stack data are being received), we can draw some final conclusions about Gen 1 technology. The maximum number of hours a Gen 1 stack accumulated without repair is 2375 (Figure 6.7), which is the longest stack durability from a FCEV in normal use that we know of.

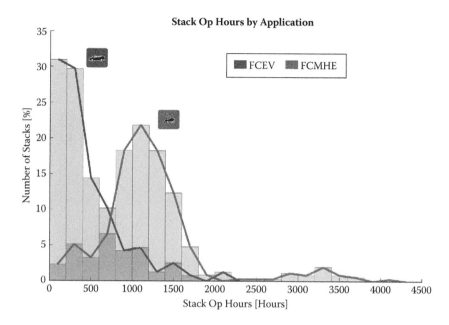

FIGURE 6.6
Fuel cell stack hours by application.

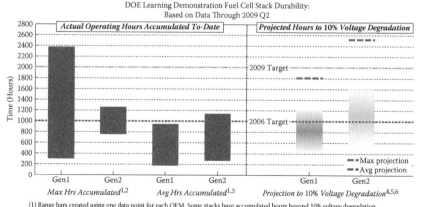

(1) Range bars created using one data point for each OEM. Some stacks have accumulated hours beyond 10% voltage degradation.
(2) Range (highest and lowest) of the maximum operating hours accumulated to-date of any OEM's individual stack in "real-world" operation.
(3) Range (highest and lowest) of the average operating hours accumulated to-date of all stacks in each OEM's fleet.
(4) Projection using on-road data -- degradation calculated at high stack current. This criterion is used for assessing progress against DOE targets, may differ from OEM's end-of-life criterion, and does not address "catastrophic" failure modes, such as membrane failure.
(5) Using one nominal projection per OEM: "Max Projection" = highest nominal projection, "Avg Projection" = average nominal projection. The shaded projection bars represents an engineering judgment of the uncertainty on the "Avg Projection" due to data and methodology limitations. Projections will change as additional data are accumulated.
(6) Projection method was modified beginning with 2009 Q2 data, includes an upper projection limit based on demonstrated op hours.

FIGURE 6.7
Stack hours accumulated and projected hours to 10% voltage degradation.

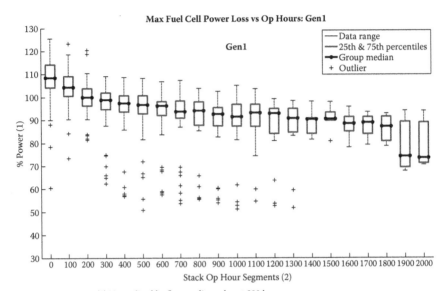

(1) Normalized by fleet median value at 200 hours.
(2) Each segment point is median FC power (+−50 hrs).
    Box not drawn if fewer than 3 points in segment.

FIGURE 6.8
Fuel cell power loss as a function of operating time for Gen 1 stacks.

On average, the slope of the initial power degradation is steeper in the first 200 hours and then becomes more gradual. We also found that around 1000 hours of data were required to reliably determine the slope of the more gradual secondary degradation (Figure 6.8). Finally, with significant drops in power observed at 1900–2000 hours, this appears to be a solid upper bound on Gen 1 stack durability (characterizing 2003–2005 technology).

For second-generation (Gen 2) fuel cell stacks (2005–2007 technology that started deployment in 2008), the maximum hours accumulated from the four teams is now approximately 800–1200+ hours; the team average hours accumulated is approximately 300–1100 hours. Relative to projected durability, the spring 2010 results indicate that the highest average projected team time to 10% voltage degradation for Gen 2 systems was 2521 hours, with a multi-team average projection of 1062 hours (Figure 6.7).

6.4.3 Hydrogen Fill Rate (Combined FCB, FCEV, FCMHE)

The hydrogen fueling rate is an important metric for many fuel cell applications. A fast fill is important to the productivity measures in the FCMHE application, because less time filling the tank means there is more time to move material. FCBs typically fuel at the end of a route and many buses may be

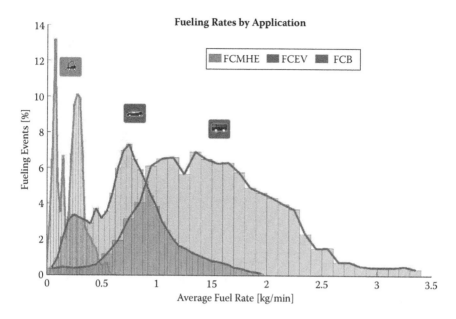

FIGURE 6.9
Hydrogen fueling rates by application.

filling at about the same time, so fast, efficient fueling is important to maintain the operation schedule. The fill time target for FCEVs is 1 kg/min and is related to consumer acceptance. Figure 6.9 combines the composite data on hydrogen fueling rates for FCMHEs, FCEVs, and FCBs. These have different hydrogen fueling rate trends: The FCBs have the highest and the MHEs have the lowest.

Many other factors such as tank capacity and level at fill influence the time to fuel. FCMHEs are slower than FCEVs and FCBs, but FCMHEs have the smallest average fill amount of about 0.5 kg/fill. FCB has the highest average fill amount of more than 20 kg/fill. The FCEV average fill amount is more than 2.1 kg/fill.

6.4.4 Fuel Economy (FCEV, FCB)

We used city and highway drive-cycle tests (Figure 6.10) on a chassis dynamometer to measure vehicle fuel economy according to draft SAE J2572 (left two bars, representing the range of four points, one from each OEM). We then adjusted these raw test results according to U.S. Environmental Protection Agency (EPA) methods to create the "window-sticker" fuel economy that consumers see when purchasing the vehicles (0.78 × Highway, 0.9 × City) (center two bars). This resulted in an adjusted fuel economy range of 42–57 miles/kg hydrogen for the four teams for Gen 1 vehicles, compared to 43–58 miles/kg

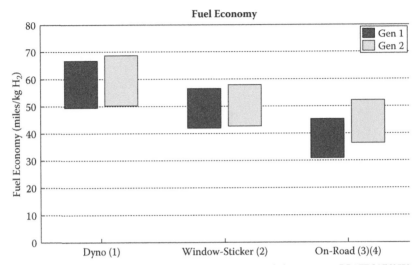

(1) One data point for each make/model. Combined City/Hwy fuel economy per DRAFT SAE J2572.
(2) Adjusted combined City/Hwy fuel economy (0.78 × Hwy, 0.9 × City).
(3) Excludes trips < 1 mile. One data point for on-road fleet average of each make/model.
(4) Calculated from on-road fuel cell stack current or mass flow readings.

FIGURE 6.10
FCEV fuel economy in miles/kg hydrogen.

for Gen 2 vehicles. As with all vehicles sold today, actual on-road fuel economy is slightly lower than this rated fuel economy (right two bars). The on-road fuel economy range is 31–45 miles/kg hydrogen for Gen 1 and 36–52 miles/kg hydrogen for Gen 2. This last comparison shows an important finding, which is that Gen 2 vehicles are considerably more robust and enable higher fuel economy (relative to Gen 1), even when they are driven under varying conditions.

The EPA has adjusted its testing and reporting methodology, beginning with model year 2008 vehicles, to make the window-sticker fuel economy better reflect on-road driving performance, but this project uses the EPA adjustment that was in place when the vehicles were introduced to avoid retesting or applying the new corrections that have not been validated for application to hydrogen FCEVs.

Figure 6.11 shows the fuel economy in diesel energy equivalent gallons for the FCBs and baseline buses we evaluated at three locations. The FCBs showed fuel economy improvement ranging from 53% to 141% compared to diesel and compressed natural gas (CNG) baseline buses. AC Transit FCBs have an overall fuel economy 53% higher than its diesel buses.* For all revenue service

---

* Because the data collection on AC Transit's diesel baseline buses was completed previously, the chart includes the average fuel economy for one year of service.

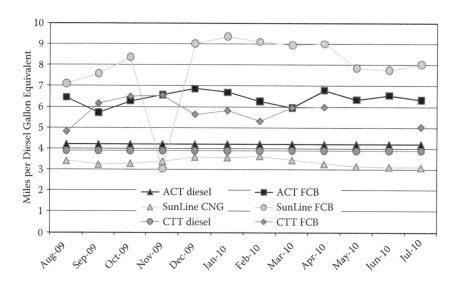

FIGURE 6.11
Fuel economy for FCBs and baseline buses.

at AC Transit, the FCBs had a fuel economy average of 62% higher than the diesel buses. During the evaluation, the FCBs reached more than twice the fuel economy of the diesel buses, even though the FCBs have air-conditioning and the diesel buses do not. In the next AC Transit demonstration and evaluation, the diesel baseline buses will include air-conditioning for a more accurate comparison. SunLine's FCB has a fuel economy 149% higher than its CNG buses. CTTRANSIT's FCB has a fuel economy 44% higher than its diesel buses. The CTTRANSIT diesel buses operate at twice the average speed as the FCB operating on the Star Route,* which causes significantly lower fuel economy than the FCBs at the other two transit agencies.

6.4.5 Range (FCEV, FCMHE)

In FY 2008, the driving range of the project's FCEVs was evaluated based on fuel economy from dynamometer testing (EPA adjusted) and onboard hydrogen storage amounts and compared to the 250-mile target. The resulting Gen 2 vehicle driving range was 196–254 miles (Figure 6.12) from the four teams, and met the 250-mile range objective. In June 2009, an on-road driving range evaluation was performed in collaboration with Toyota and Savannah River National Laboratory. The results indicated that a 431-mile on-road range was possible in southern California using Toyota's FCHV-adv fuel cell vehicle (3).

---

* CTTRANSIT operates its FCB on a downtown shuttle route—the Star Route—which is characterized by slow speeds, multiple stops, and longer idle times.

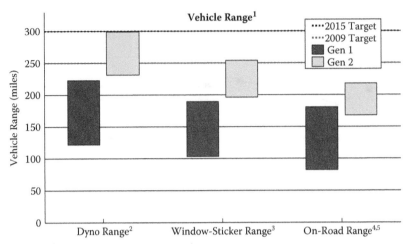

[1]Range is based on fuel economy and usable hydrogen on-board the vehicle.
    One data point for each make/model.
[2]Fuel economy from unadjusted combined City/Hwy per DRAFT SAE J2572.
[3]Fuel economy from EPA Adjusted combined City/Hwy (0.78 × Hwy, 0.9 × City).
[4]Excludes trips < 1 mile. One data point for on-road fleet average of each make/model.
[5]Fuel economy calculated from on-road fuel cell stack current or mass flow readings.

FIGURE 6.12
FCEV range in miles.

More recently, the significant on-road data we obtained from Gen 1 and Gen 2 vehicles enabled a comparison of the real-world driving ranges of all the vehicles in the project. The data show a 63% improvement in the median real-world driving range of Gen 2 vehicles (91 miles) compared to Gen 1 (56 miles), based on actual distances driven between nearly 29,000 fueling events (Figure 6.13). The vehicles are capable of two to three times greater range, but the median distance traveled between fuelings is one way to measure the improvement in the vehicles' capability and the way they are actually being driven.

FCMHEs do not measure range in miles but can use other data metrics such operation times between fuelings. Figure 6.14 and Figure 6.15 show the operation times between fuelings for the ARRA-funded FCMHEs and the DLA-funded FCMHEs. The average operation time between fuelings was 3.8–4.1 hours. This represents actual use and not the maximum capabilities of the systems that include Class 1, 2, and 3 trucks.

One factor that affects the time between fuelings is an opportunity fill or a fill that is completed because of convenience rather than because the tank was almost empty. Another factor is the average daily hours per fleet (2.2–7 hours). This is an example of a trend that relies on a number of data metrics. We will study this over the next year, along with other operation trends that are site and operator dependent.

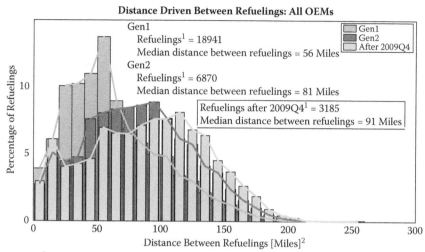

FIGURE 6.13
FCEV miles between fuelings.

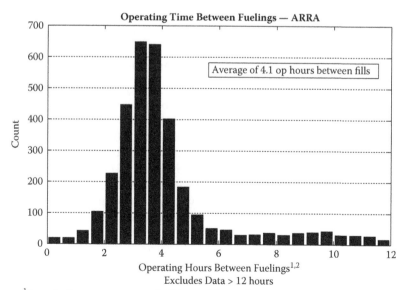

FIGURE 6.14
ARRA FCMHE operating time between fuelings.

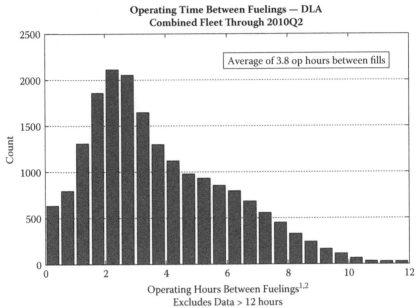

FIGURE 6.15
DLA FCMHE operating time between fuelings.

6.4.6 Hydrogen Produced and/or Consumed (Combined FCEV, FCMHE)

As of the end of June 2010, 134,395 kg and 19,831 kg were produced or consumed, or both, by FCEVs and FCMHEs respectively. Figure 6.16 shows the cumulative hydrogen consumed by FCEVs and FCMHEs over time. The FCEVs were operational and supplying data in 2005, and the consumed hydrogen amount increased steadily until the end of 2009, when some of the projects were ending. The first set of operational FCMHEs was in 2009; in 2010 many more FCMHEs were deployed. We thus see an increase in the hydrogen consumed. The amount of hydrogen consumed should increase quickly with hundreds of FCMHE units in operation.

6.4.7 Efficiency (FCEV)

Automotive company researchers measured fuel cell system efficiency from select vehicles on a vehicle chassis dynamometer at several steady-state points of operation. NREL helped to ensure that appropriate balance-of-plant electrical loads were included. This ensured the results were comparable to the target

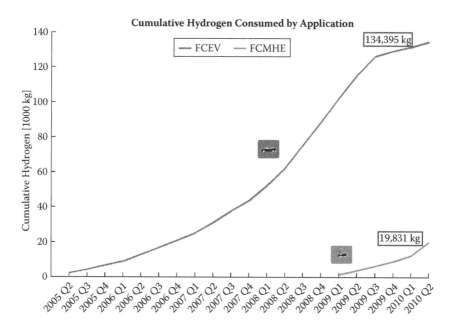

FIGURE 6.16
Cumulative hydrogen consumed by application.

and based on the entire system rather than only the stack. DOE's technical target for net system efficiency at quarter-power is 60%. Baseline data from the four Learning Demonstration teams several years ago showed a net system efficiency of 51%–58% for Gen 1 systems, which was very close to the target. As Gen 2 vehicles were introduced, the companies also performed baseline dynamometer testing that revealed an efficiency of 53%–59% at quarter-power, within one percentage point of the target. Since the last progress report was published, we have also expanded this CDP to include a comparison of the efficiency at full power, where DOE's target was 50% net system efficiency (Figure 6.17). The data show Gen 1 systems have 30%–54% efficiency at full power; Gen 2 systems have 42%–53% efficiency, exceeding the 50% target. We published the ranges of efficiency data from the four teams with the two shaded green sections, showing that Gen 2 data are more closely clustered than Gen 1 data.

6.4.8 Reliability (FCB, FCBU)

FTA requirements for 40-ft diesel bus life are 12 years or 500,000 miles. Transit agencies typically keep these buses for as long as 14 years, rebuilding the engines at approximately midlife. To match this durability, a fuel cell power system should be able to operate for half the life of the bus. FTA has

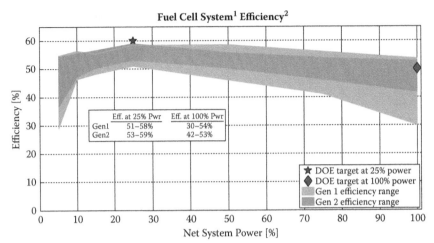

Fuel Cell System[1] Efficiency[2]

[1]Gross stack power minus fuel cell system auxiliaries, per DRAFT SAE J2615.
  Excludes power electronics and electric drive.
[2]Ratio of DC output energy to the lower heating value of the input fuel (hydrogen).
[3]Individual test data linearly interpolated at 5,10,15,25,50,75, and 100% of max net power.
  Values at high power linearly extrapolated due to steady state dynamometer cooling limitations.

FIGURE 6.17
Fuel cell system efficiency as a function of power.

set an early performance target of 4–6 years (or 20,000–30,000 hours) durability for the fuel cell propulsion system.

Figure 6.18 illustrates that demonstrated reliability increases over time for these fuel cell systems. Tracking the transit industry measure of reliability, the blue line shows the monthly average miles between road calls (MBRCs) for all five buses (fuel cell system only). These data show a significant increase in fuel cell–related MBRCs after the new fuel cell systems were installed. (The shaded area marks the timing of the fuel cell power system installations.) Overall reliability for the fuel cell system has increased by 41% since the new version was installed in early 2008. The black dotted line (trailing 12-month average) clearly shows the upward trend over time.

The backup power market requires a reliable backup power method. We are analyzing this specific topic. Figure 6.19 shows the number of successful and attempted starts by month of operation from August 2009 to June 2010.

During this period 201 starts were attempted, of which 199 were successful, for approximately 99% reliability. FCBU units may have infrequent operation. Starts classified as conditioning are also identified and make up about 57% of the starts. A condition start is an automated short operation for regular system checks and is initiated after long periods of inactivity.

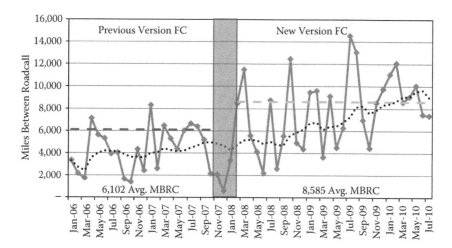

FIGURE 6.18
Average monthly MBRC for the fuel cell power system.

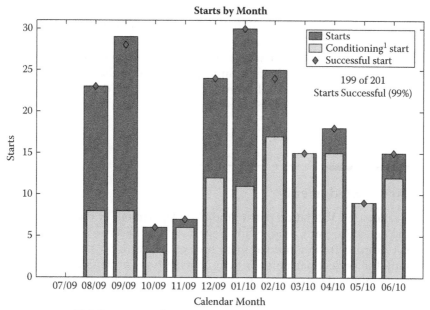

(1) FC system conditioning is an automated operation for regular
system checks; activitated after long periods of no operation.

FIGURE 6.19
FCBU starts by month.

6.5 Conclusions

The HSDC has analyzed more than 470 proton exchange membrane (PEM) fuel cell systems in light-duty automotive, heavy-duty transit, material handling, and backup power applications. We expect to analyze more than 800 in the coming years. The systems have accumulated more than 410,000 hours in real-world settings and 190,000 kg of hydrogen has been produced or consumed, or both. The HSDC does not analyze all the fuel cell applications and deployed units, but the analyzed systems do provide a well-rounded status on the technology. We have been able to apply and transform our FCEV analysis techniques to many other applications and found that some data topics can be applied across all the applications and other data topics that are application specific. For example, durability is less important in a backup power application than in other applications because backup power units do not operate many hours in a year.

Our analysis evolves as we identify data trends and accumulate more data and collaborations with industry and stakeholders. Many analysis topics, such as analysis of performance metrics that affect the application market value proposition, will be added. We have observed many metrics, such as efficiency and emissions, where these fuel cell systems are competitive in certain applications now. We have also studied metrics such as durability and cost where additional research and development are needed.

With all these deployments, many lessons have been learned that support technology advancements, market penetration, education, and development of end users who are comfortable and knowledgeable about the many benefits of fuel cell technologies: low emissions, decreased dependence on foreign fossil fuels, reduced noise, increased efficiency, and ease of use.

6.6 Acknowledgments

The Technology Validation part of the U.S. Department of Energy Fuel Cell Technologies Program, along with Department of Defense and the Federal Transit Administration, funded these projects. In addition, the authors wish to thank the project partners for their contributions of detailed raw data to NREL as well as their valuable feedback on our methodologies and results.

References

1. Eudy, L., Chandler, K., Gikakis, C., Fuel Cell Buses in U.S. Transit Fleets: Current Status 2010, October 2010 NREL/TP -560-49379 (October 2010).
2. K. Wipke, S. Sprik, J. Kurtz, T. Ramsden, Learning Demonstration Interim Progress Report, July 2010 NREL/TP-560-49129 (September 2010).
3. Wipke, K., Anton, D., and Sprik, S., Evaluation of Range Estimates for Toyota FCHV-adv under Open Road Driving Conditions, prepared under SRNS CRADA number CR-04-003, August 2009.
4. Kurtz, J., Wipke, K., Sprik, S., and Ramsden, T. Early Fuel Cell Market Deployments: ARRA, August 2010 NREL/TP-5600-49602 (August 2010).
5. Kurtz, J., Wipke, K., Sprik, S., and Ramsden, T. Fall 2010 Composite Data Products ARRA Material Handling Equipment, September 2010 NREL/TP-5600-49603 (September 2010).
6. http://www.nrel.gov/hydrogen/proj_tech_validation.html.

7

Producing Hydrogen for Vehicles via Fuel Cell–Based Combined Heat, Hydrogen, and Power: Factors Affecting Energy Use, Greenhouse Gas Emissions, and Cost

Darlene Steward, Karen Webster, and Jarett Zuboy

CONTENTS

7.1 Introduction

This chapter introduces the concept of producing fuel for hydrogen-powered vehicles using combined heat, hydrogen, and power (CHHP) systems based on stationary high-temperature fuel cells, which also provide electricity and heat to buildings. In addition, it explores the factors affecting the performance of CHHP systems in various locations as well as the associated greenhouse gas (GHG) emissions and hydrogen cost.

Compared with the current road transportation system based on gasoline- and diesel-powered vehicles, a system based on hydrogen-powered vehicles offers important environmental and energy security benefits. These include eliminating tailpipe emissions that create local air pollution, reducing life cycle pollutant and greenhouse gas emissions, and drawing on domestic primary energy sources. In addition, there are opportunities to leverage the environmental and energy benefits of vehicular hydrogen and fuel cells with stationary power applications and renewable energy generation. Research, development, and deployment programs over the past several decades have reduced the barriers to widespread implementation of fuel cell technologies, hydrogen-powered vehicles, and hydrogen fueling infrastructure, but some barriers remain.

One of the key barriers to the commercialization of hydrogen vehicles is the lack of hydrogen refueling infrastructure. Though proven technologies exist for delivering hydrogen to stations or producing hydrogen at stations, such as on-site steam methane reforming or electrolysis, stakeholder engagement proves challenging due to the high capital costs of refueling stations and uncertainty around future demand for hydrogen vehicles and therefore hydrogen fuel (Greene et al. 2008; NRC 2008). One of the complicating requirements for infrastructure rollout is the need for small stations to serve early vehicle markets. Early stations will likely be underused as local hydrogen fuel cell electric vehicle (FCEV) fleets may increase in size only slowly over time. The uncertain rate of growth in demand from FCEVs increases the financial risks associated with large stations, which are capital intensive. Deployment of smaller stations also results in a larger number of refueling locations serving early vehicle markets for a given capital outlay, stimulating demand for more FCEVs by removing the consumer adoption barrier of limited refueling availability (Melaina 2003; Melaina and Bremson 2006). Small stations also address hesitancy on the part of key stakeholders to make large commitments to novel technological systems perceived as having high technical risk (IPHE 2010). Convenient, low-cost, and small-scale hydrogen refueling can therefore facilitate the investment process and accelerate vehicle adoption during the transition phase of hydrogen infrastructure development.

One way to increase the availability of small hydrogen refueling stations is to produce excess hydrogen from high-temperature fuel cells installed in combined heat and power (CHP) configurations. CHP applications are an established technology capable of utilizing energy resources more efficiently than conventional centralized electricity systems while reducing overall emissions. A typical combined efficiency for electricity from centralized power plants and heat generation in natural gas boilers is 45%, compared to up to 80% for a comparable CHP system (Shipley et al. 2008). Carbon dioxide emissions from CHP systems can be half those from traditional systems, depending upon the source of the electricity being displaced (ICF International 2008). Of the 85 GW of CHP installed in the United States, providing about 12% of all electricity, 88% is in industrial applications and 12% is in commercial and institutional building applications. The majority of CHP applications utilize reciprocating engines (46% of facilities), and the majority of installed CHP capacity is large combined cycle natural gas turbines (53% of total installed capacity) (EEA 2007).

Combining hydrogen production with CHP systems results in CHHP or trigeneration systems. These systems would benefit from sustained operation and known savings associated with heat and power while also providing hydrogen at low volumes. We focus on building applications where hydrogen fueling capability may be advantageous for either private FCEV fleets ("behind the fence") or through a public outlet. If the building location is not suitable as a refueling location, hydrogen delivery costs can be relatively small if installations are located near suitable retail locations. In CHHP applications co-located with centrally controlled vehicle fleets, financial risks can be reduced compared to public retail hydrogen stations due to increased certainty around future demand. These installations therefore may involve multiple stakeholder groups, including building managers, utilities, and one or more fleet managers. In addition to facilitating early hydrogen infrastructure development, the CHHP strategy may also result in a durable, long-term market outside the public light-duty vehicle sector, especially in cases where buildings or building owners have or are partnered with "captive" hydrogen demand from forklifts, ground equipment, buses, or hydrogen vehicles.

Figure 7.1 provides a simple depiction of a CHHP stationary fuel cell system. The upper portion of the schematic indicates the baseline or conventional means of supplying a building with electricity from the grid and heat for space and water heating from natural gas combusted in a boiler. The lower portion of the schematic shows a stationary fuel cell converting natural gas into electricity and heat for the building as well as a side stream of hydrogen for use in vehicular applications. A variety of distributed generation technologies can be deployed exclusively in CHP applications, including reciprocating engines, gas turbines, steam turbines (given an appropriate source of steam), and micro

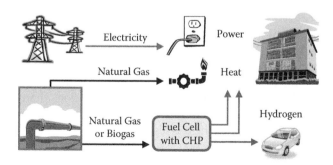

FIGURE 7.1
Schematic of the combined heat, hydrogen, and power energy system.

turbines.* Fuel cells are unique among these distributed energy technologies in their capability to produce excess hydrogen in addition to heat and power. Some fuel cells produce hydrogen through external reforming of methane, including polymer electrolyte membrane (PEM) and phosphoric acid fuel cells (PAFC), while other higher-temperature systems produce hydrogen internally, including solid oxide (SOFC) and molten carbonate fuel cells (MCFC). Molten carbonate fuel cells are modeled in this study.

CHHP systems are an emerging technology, and few real-world performance and cost data are available. This chapter explores, via modeling and analysis, the various factors that affect the potential commercial viability and energy/greenhouse gas emissions benefits of CHHP based on different geographic locations, building applications, and fuel cell system sizes.

Sections 7.2 and 7.3 describe the CHHP and baseline systems, our modeling methods, and the factors affecting CHHP energy use, GHG emissions, and hydrogen cost compared with the baseline system. Our analysis approach assumes that there is hydrogen demand in a specific location, and the viability of CHHP hydrogen production from a specific building is being considered in comparison to the viability of small-scale steam methane reforming (SMR) hydrogen production at or near that location.

The remainder of the chapter summarizes the geographic, economic, and system conditions that tend to result in the lowest CHHP-based hydrogen costs and the largest energy and GHG-reduction benefits, as well as the potential for CHHP-powered buildings to provide hydrogen to satisfy demand from early FCEVs. In addition to comparing CHHP versus baseline results, we make some comparisons across locations and building types to examine the types of buildings and locations that might be most favorable for CHHP hydrogen production. Our analysis results suggest that buildings with relatively uniform electricity and heating loads throughout the day and night in climates with little seasonal variation are the most economically favorable for fuel cell CHHP installations. Low natural gas prices reduce costs while low grid electricity prices are not favorable because the electricity generated by the fuel cell is less valuable in competition with low electricity prices. Total GHG emissions are the sum of on-site emissions associated with combustion/reforming of natural gas and upstream emissions associated with generation of electricity and, to a much lesser extent, production of natural gas. As a result, the GHG emissions benefit (or penalty) for installation of the CHHP system can be expressed as the ratio between the emissions savings from substituting fuel cell–generated electricity for grid electricity and the emissions penalty for increased natural gas usage. In all the cases presented here, the CHHP system had lower GHG emissions than the baseline system, although the magnitude of the benefit was highly dependent on the regional electricity generation mix. It is also worthwhile to note that our analysis did not include the GHG reduction associated with substitution of hydrogen

---

* The U.S. Environmental Protection Agency provides a catalog of CHP technologies (http://www.epa.gov/chp/basic/catalog.html).

for gasoline in vehicles, which would have greatly increased the reported GHG benefit. Although quantitative results are presented, trends related to the effects various factors have on CHHP costs and benefits are the intended focus of the analysis. Because of the pre-commercial nature of CHHP technology, substantial uncertainty exists with regard to quantitative cost and performance estimates.

7.2 Modeling Fuel Cell CHHP Systems with the FCPower Model

This section describes the methodology we used to model the technical and economic performance of fuel cell–based CHHP systems. Our analytical tool is the Fuel Cell Power (FCPower) model, which was developed for the U.S. Department of Energy by the National Renewable Energy Laboratory. The FCPower model analyzes CHHP system performance over an entire year, including 8760 hourly time steps, with the fuel cell responding to specified building demands for electricity and economic calculations taking into account fuel prices, initial capital and maintenance costs over time, and financial and tax considerations. It uses user-defined inputs, default values and calculations, and a standard discounted cash flow rate of return methodology to determine the cost of delivered energy from a CHHP system, with reference to a specified after-tax internal rate of return. It also determines the amount and type of energy input and output and the associated GHG emissions. The FCPower model is publicly available for download as an Excel spreadsheet, along with detailed supporting documentation (DOE 2011).

7.2.1 Configuration and Performance of the Modeled Molten Carbonate Fuel Cell CHHP System

The FCPower model can analyze CHHP systems based on MCFCs, PAFCs, and SOFCs. However, fuel cells have just begun to be used in CHP applications, and CHHP hydrogen production has been demonstrated in the field only for MCFCs. Therefore, the modeling described in this chapter is based on an MCFC CHHP system, and much of the performance information is derived from modeling and performance characteristics of fuel cell CHP systems rather than CHHP systems.

Figure 7.2 is a schematic of the MCFC CHHP configuration used in the FCPower model. The system concept was modeled with ASPEN Plus, a steady-state thermodynamics simulation software, using conventional industrial unit operations integrated into a novel system. This detailed model was used to create a simplified linear model of system performance within the FCPower Model framework so that FCPower results approximate the ASPEN results within a reasonable range of system performance. A detailed ASPEN process flow diagram and accompanying energy and material flows are provided in the FCPower model documentation (DOE 2011).

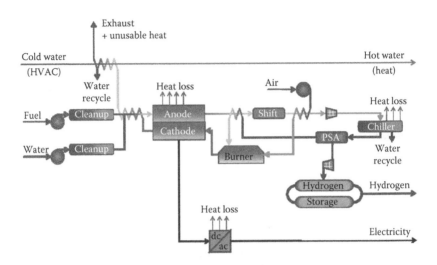

FIGURE 7.2
Schematic of the modeled MCFC CHHP system. (Courtesy of Michael Penev, NREL.)

Unlike alkaline, phosphoric acid, and polymer electrolyte membrane fuel cells, MCFCs do not require an external reformer to convert fuels to hydrogen. Owing to the high operating temperature, fuels are converted to hydrogen within the fuel cell itself via SMR, which reduces system complexity. In addition, MCFCs are not prone to carbon monoxide (CO) poisoning; CO is used as fuel along with hydrogen (H_2). Molten carbonate fuel cells can be more efficient than PAFCs, with efficiencies approaching 50%, compared with 37%–42% for PAFCs. For MCFC configurations in which waste heat is used for additional electricity generation, electrical efficiencies greater than 60% are possible, and overall fuel efficiencies can be as high as 85% (DOE 2011).

The primary disadvantage of current MCFC technology is durability. The high temperatures at which these cells operate, and the corrosive electrolyte used, accelerate component breakdown and corrosion, decreasing cell life. Corrosion-resistant component materials are being developed, along with fuel cell designs that increase cell life without decreasing performance. Another disadvantage of MCFCs is their slow responsiveness to demand fluctuations. Because the system must balance the fuel cell temperature while maintaining an even temperature distribution, long times are required for the temperature to distribute as the output level changes.

The following description of MCFC operation is taken from the *Fuel Cell Handbook, 7th edition* (EG&G 2005), which should be referenced for additional technology details. These are the half-cell electrochemical reactions:

$$H_2 + CO_3^{2-} \rightarrow H_2O + CO_2 + 2e^- \ (anode) \tag{7.1}$$

$$\frac{1}{2}O_2 + CO_2 + 2e^- \rightarrow CO_3^{2-} \ (cathode) \tag{7.2}$$

The following is the overall cell reaction:

$$H_2 + \frac{1}{2}O_2 + CO_2 \rightarrow H_2O + CO_2 \tag{7.3}$$

Besides the reaction involving H_2 and O_2 to produce H_2O, this equation shows a transfer of CO_2 from the cathode gas stream to the anode gas stream via the CO_3^{2-} ion. The need for CO_2 at the cathode requires that either CO_2 is transferred from the anode exit gas to the cathode inlet gas, CO_2 is produced by combusting the anode exhaust gas (which is mixed directly with the cathode inlet gas), or CO_2 is supplied from an alternate source. It is usual practice in an MCFC system that the CO_2 generated at the anode (right side of Equation 7.1) be routed (external to the cell) to the cathode (left side of Equation 7.2).

Hydrogen production can be increased by using useful heat from the electricity production process. Hydrogen production is a means of cooling the fuel cell, and available useful heat is inversely proportional to hydrogen output (Figure 7.3). Owing to this direct relationship between hydrogen production and heat generation, more hydrogen can be produced at higher electricity output levels, although the electrical efficiency of the fuel cell is slightly reduced. Table 7.1 summarizes the operating parameters underlying these output dynamics, with efficiencies indicated for maximum electricity output.

The MCFC system represented in the FCPower model is electricity load following, meaning that the fuel cell automatically ramps up and down within its operational range in response to the electricity load. It operates in two modes. In the hydrogen production mode, 70% of the caloric content of the fuel mixture entering the anode is used to make electricity (as in a CHP

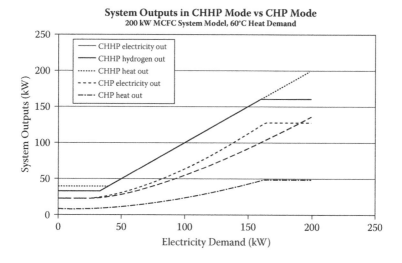

FIGURE 7.3
Variations in CHHP system outputs versus CHP system outputs as a function of electricity demand.

TABLE 7.1

Operating Parameters for a Generic MCFC System

Fuel Cell System Operating Parameters	Units	Value	Notes
Conversion Efficiencies (Max. Values)			
Electricity production efficiency	kWhe/kWh total energy consumption	41%	LHV basis, thermodynamic models
Hydrogen production efficiency	kWhH2/kWh total energy consumption	26%	
Useful heat production efficiency	kWth/kWh total energy consumption	10%	
Total system efficiency	(kWe+kWH2+kWth)/ kW total	77%	
Additional Operating Parameters			
Hydrogen production efficiency	kW H2/kW CHP heat reduced	96%	Based on thermodynamic models
Hydrogen over-production efficiency	kW H2/additional kW fuel consumed	80%	Based on thermodynamic models
Fraction of heat convertible to hydrogen (max.)	Percentage of total useful heat	65%	Based on thermodynamic models
Hydrogen overproduction fraction (max.)	Percentage of H2 production	50%	Based on thermodynamic models
Electricity response time	Percent change per hour	10%	Default value

system), and 70% of the hydrogen in the anode exhaust gas is recovered and stored. In the hydrogen overproduction mode, 60% of the caloric content of the fuel mixture entering the anode is used to make electricity, and 75% of the hydrogen in the anode exhaust gas is recovered and stored. Both modes reduce the amount of energy available for heating the building by producing excess hydrogen. In the current analysis, the fuel cell is operated in hydrogen overproduction mode to maximize the amount of hydrogen produced. Because of the low availability of heat from the system, thermally activated cooling is not considered. When the CHHP system does not provide sufficient electricity and/or heat to meet building demands, supplementary electricity and/or natural gas are purchased to meet the demands.

7.2.2 Major CHHP and Baseline System Modeling Assumptions

Our analyses assumed that suitable buildings and hydrogen demand exist in each location modeled. We also assumed that the building energy and hydrogen demands can be met using one of two systems. The CHHP system provides building heat and electricity as well as hydrogen via an MCFC, sells electricity to the grid when electricity production exceeds building demand,

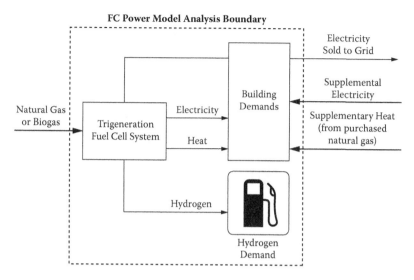

FIGURE 7.4
Boundaries and energy flows for the CHHP system.

and purchases supplementary electricity and/or natural gas when the MCFC cannot meet all building demands. The baseline system provides building electricity and heat exclusively via the electrical grid and natural gas heating and produces hydrogen from natural gas using small, on-site SMR systems.* The CHHP system produces hydrogen in proportion to the amount of electricity it generates, and the baseline system produces an equivalent amount of hydrogen via SMR. We assume that the hydrogen compression, storage, and dispensing configurations for the CHHP and baseline systems are similar.

The economics of CHHP systems are complex due to the influence of fuel cell dynamics and electricity prices on system economics and the inclusion of three revenue streams in the cost calculations: heat, hydrogen, and electricity. In Figure 7.4 (CHHP system) and Figure 7.5 (baseline system), arrows that cross the analysis boundary are explicitly accounted for in the FCPower model's discounted cash flow analysis. Internal arrows represent the avoided costs (revenue) from supplying electricity, heat, and hydrogen via the fuel cell system. The model solves for the total revenue derived for these energy services that would be required to equal to the annualized profited cost for the fuel cell installation. The model solves for a total cost of energy in dollars per kWh without explicitly distinguishing between the three energy products (electricity, heat, and hydrogen). The model's user specifies prices for two of the three energy products, and these values are subtracted from the total energy cost to calculate the revenue (or avoided cost) derived from the third energy product.

---

* In the current study, commercial (non-renewable) natural gas is the assumed feedstock, although biogas is also being considered for stationary CHP and CHHP applications (DOE 2009), and the FCPower model is capable of analyzing systems using this feedstock.

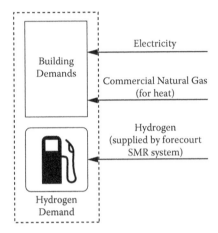

FIGURE 7.5
Boundaries and energy flows for the baseline system.

For the analyses described in this chapter, electricity supplied by the fuel cell was assigned the same value as electricity purchased from the grid, and heat from the fuel cell was assigned the same value as would have been paid for heating from a natural gas heating system. Hydrogen, by default, was the third (free) variable; thus the remaining energy costs were allocated to hydrogen. In other words, we set the values of electricity and heat generated by and used in the CHHP systems to be equal to the values of the electricity and heat used in the baseline systems, so the remaining CHHP energy costs were allocated to hydrogen production. Therefore, we can compare the CHHP hydrogen cost with the cost of hydrogen from the baseline SMR system.

The cost of stand-alone SMR-produced hydrogen decreases with increased hydrogen production (Figure 7.6) because of assumed economies of scale associated with building a larger facility. The SMR cost curve was developed using the current forecourt SMR system modeled using the H2A model (DOE 2011).[*] The plant was scaled to a design capacity of 700 kg/day using an engineering scaling factor of 0.6 for the capital equipment and default scaling assumptions for other costs. Plant output below 700 kg/day was obtained by curtailing the operation of the plant. The SMR unit is assumed to operate at an efficiency of 68.5%, including hydrogen purification and compression, with 94% of its energy supplied with natural gas and 6% supplied with electricity, primarily for compression of the hydrogen.[†] This curve shows that low-volume production of hydrogen using SMR is likely to be expensive, which is a justification for exploring the use of CHHP-based hydrogen production. Because CHHP systems are in an early stage of development, their

---

[*] File name Current_Forecourt_Hydrogen_Production_from_Natural_Gas_(1,500_kg_per_ day)_v2.1.2.xls.

[†] H2A Current Forecourt Hydrogen Production from Natural Gas (1,500 kg per day) version 2.1.2 http://www.hydrogen.energy.gov/h2a_prod_studies.html).

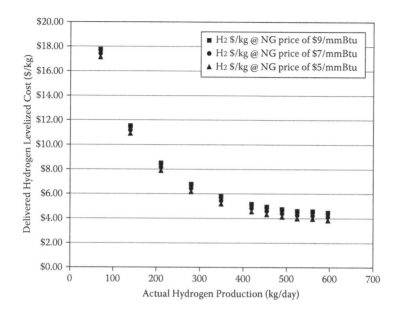

FIGURE 7.6
Delivered hydrogen cost from the SMR system at various production levels and natural gas costs, from H2A model (DOE, DOE Fuel Cell Power Analysis, 2011).

costs are not as well established as SMR costs, and cost estimates depend on uncertain performance and economic assumptions. Subsequent sections of this chapter describe trends in CHHP performance and cost as they vary under different operating conditions. For most energy use and GHG emissions comparisons, the fuel cell is sized at the average building electrical load. Refueling station assumptions for the CHHP system are similar to those for the H2A forecourt model, which was used as the basis for the SMR system in the baseline analysis (i.e., the CHHP system station assumptions were matched as closely as possible to the baseline system). We assumed a CHHP system life of 20 years to match the assumed SMR system life. The CHHP modeling includes replacement of fuel cell stacks and other equipment during the system lifetime. We also assumed an uninstalled direct capital cost of $2200/kW. This capital cost estimate was based on estimates of MCFC CHP capital costs at high-volume production levels (at least 100 MW/yr) rather than current, low-volume costs (Figure 7.7). Detailed assumptions are provided in the FCPower model and its documentation (DOE 2011).

7.2.3 Geographic and Building-Type Considerations

In addition to the economic and performance assumptions listed above, the analyses in this chapter are based on geographic and building data developed by the National Renewable Energy Laboratory's Electricity, Resources, and Building Systems Integration Center for the U.S. Department of Energy's

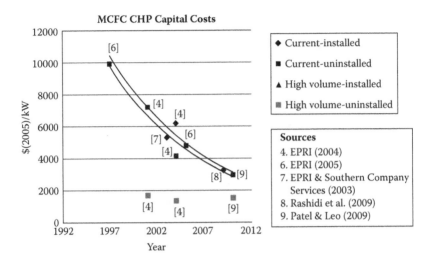

FIGURE 7.7
Collected cost information for MCFC CHP plants. (Courtesy of Genevieve Saur, NREL.)

Building Technologies Program (Field et al. 2010). The data include electricity and heat load profiles for various building types in U.S. cities across eight climate zones (Figure 7.8). Hospitals, large office buildings, and large hotels were evaluated for this study because they are large energy users that could potentially benefit from using CHHP systems (Steward 2009). The following cities were evaluated because they represent various climate zones and have potential for high hydrogen demand:

- Miami (zone 1)
- Houston (zone 2)
- Los Angeles (zone 3)
- Baltimore (zone 4)
- Chicago (zone 5)
- Minneapolis (zone 6)

Another factor that varies by geographic location is the GHG intensity of electricity generation. Different parts of the country use different mixes of fuels (coal, natural gas, nuclear, petroleum, and renewables) to generate electricity, and these mixes emit different amounts of GHGs per unit of electricity generated. Table 7.2 shows these GHG intensities, or emissions factors, for cities in each of the climate zones. The emissions factors account for GHG emissions due to electricity generation and transmission losses (EPA 2010).

Upstream and on-site GHG emissions are also produced by the natural gas used for heating the building and producing hydrogen in the baseline system

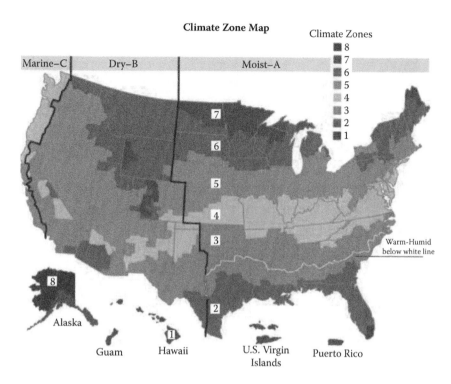

FIGURE 7.8

U.S. climate zones (DOE, DOE Fuel Cell Power Analysis, 2011).

TABLE 7.2

Regional Grid Electricity Emissions Factors for
Representative Cities in the Eight Climate Zones

Climate Region	City	Grid Electricity Emissions Factor (g CO_2eq/kWh Available at Outlet)
1	Miami	654
2	Houston	693
3	Los Angeles	409
4	Baltimore	733
5	Chicago	568
6	Minneapolis	768
7 & 8	Fairbanks	631

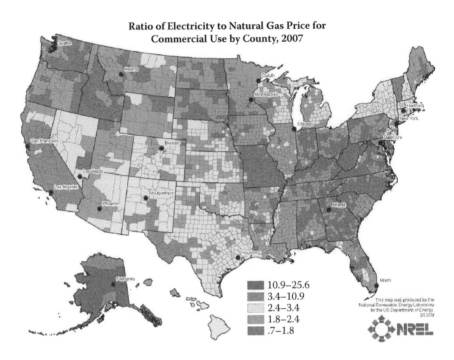

Ratio of Electricity to Natural Gas Price for Commercial Use by County, 2007

10.9–25.6
3.4–10.9
2.4–3.4
1.8–2.4
.7–1.8

FIGURE 7.9
Ratio of electricity to natural gas price for commercial use by county, 2007.

as well as operating the fuel cell and providing supplementary heat in the CHHP system. However, these emissions do not vary by location in our analyses. Upstream emissions associated with extraction and transport of the natural gas are assumed to be 35 g CO_2eq/kWh of fuel energy content (GREET 1.8d).* On-site emissions from natural gas use are based on the carbon content of the fuel and are assumed to be 209 g CO_2eq/kWh of fuel energy content.

The cost analysis in this study accounts for regional variations in electricity and natural gas prices. Natural gas prices directly impact the economics of both the CHHP system and the baseline system. Higher natural gas prices result in higher costs for both systems. Electricity prices also have a direct impact on the baseline system and the supplementary electricity that must be purchased for the CHHP system. However, the benefit derived from producing electricity from the fuel cell rather than buying it from the grid increases with increasing grid electricity price, making the fuel cell system more economically favorable in comparison to the baseline system. The ratio of the price of delivered energy for electricity to natural gas provides an estimate of the economic benefit. The ratio of electricity to natural gas price ("spark spread") is mapped by county in Figure 7.9. A higher ratio indicates more

---

* The Greenhouse Gases, Regulated Emissions, and Energy Use in Transportation Model (http://greet.es.anl.gov/).

favorable economics because low natural gas prices reduce CHHP costs and high electricity prices increase the value of CHHP generated electricity.

For the analyses presented here, current regional electricity and natural gas prices were obtained from the U.S. Energy Information Administration.* Table 7.3 summarizes the climate and energy costs for the cities included in this chapter. Three of the cities listed in Table 7.3—Miami, Los Angeles, and Chicago—were selected for more in-depth analysis throughout the chapter. These three cities were chosen because they represent a range of climate conditions, grid GHG emissions profiles, and energy costs found in the United States, as well as being relatively large metropolitan areas that might be early markets for hydrogen.

7.3 Functional Characteristics That Affect CHHP System Performance

As mentioned previously, CHHP systems are an emerging technology, and few real-world performance and cost data are available. However, the functional characteristics that affect the performance of CHHP systems have been identified and modeled. This section explores those functional characteristics. The next section explores how real-world conditions affect the functional characteristics and, consequently, CHHP system performance.

The output of the MCFC CHHP system that we are modeling increases and decreases in response to the electricity demand of the building in which it is installed (see Section 7.2). When building electricity demand is at or above the fuel cell's capacity, the fuel cell operates at 100% capacity. At this level, the amount of heat available for building applications and hydrogen production is maximized. In general, the electrical efficiency of the CHHP system is higher than the efficiency of grid-delivered electricity, and utilizing CHHP-generated heat in the building increases the total efficiency of the system. Therefore, increasing the amount of CHHP-generated electricity and heat used to meet the building's demands decreases energy consumption compared with using conventional energy sources (grid electricity and natural gas heating).

An ideal CHHP building application would have a constant electricity demand that is met entirely by the fuel cell while requiring the fuel cell to operate at 100% capacity at all times (i.e., the fuel cell would be sized to meet all demand operating at full capacity). In addition, the building's heat demand would be met by the CHHP system while consuming all of the fuel cell's useful

---

* Natural gas: http://www.eia.gov/dnav/ng/ng_pri_sum_a_EPG0_PCS_DMcf_a. htm; Price (in $2005): 7.92 $/MMBTU.
  Electricity: http://www.eia.doe.gov/cneaf/electricity/epm/table5_3.html; Price (in $2005): 10.26 $/kWh. 2010 averages, accessed 3/25/2011.

TABLE 7.3

Summary of CHHP and Baseline Systems by Location and Building Type

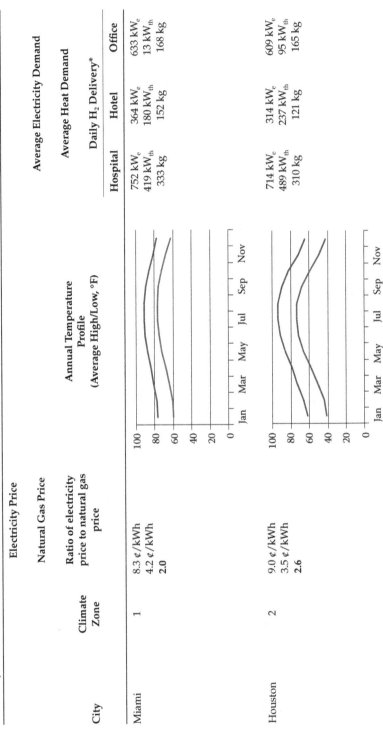

City	Climate Zone	Electricity Price / Natural Gas Price / Ratio of electricity price to natural gas price	Annual Temperature Profile (Average High/Low, °F)	Average Electricity Demand / Average Heat Demand / Daily H₂ Delivery*		
				Hospital	Hotel	Office
Miami	1	8.3 ¢/kWh 4.2 ¢/kWh 2.0		752 kW$_e$ 419 kW$_{th}$ 333 kg	364 kW$_e$ 180 kW$_{th}$ 152 kg	633 kW$_e$ 13 kW$_{th}$ 168 kg
Houston	2	9.0 ¢/kWh 3.5 ¢/kWh 2.6		714 kW$_e$ 489 kW$_{th}$ 310 kg	314 kW$_e$ 237 kW$_{th}$ 121 kg	609 kW$_e$ 95 kW$_{th}$ 165 kg

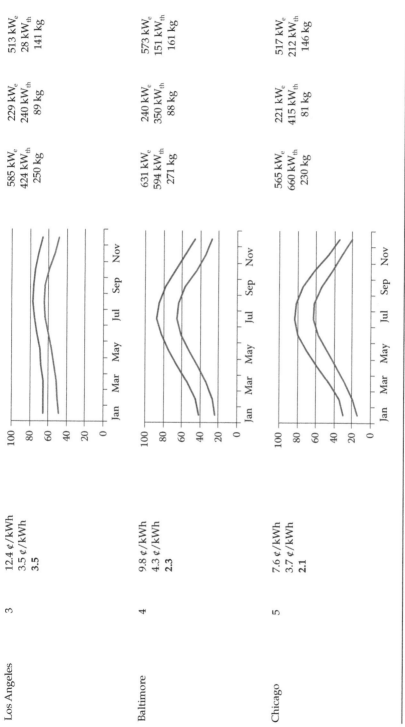

(continued)

TABLE 7.3 (CONTINUED)

Summary of CHHP and Baseline Systems by Location and Building Type

City	Climate Zone	Electricity Price / Natural Gas Price / Ratio of electricity price to natural gas price	Annual Temperature Profile (Average High/Low, °F)	Average Electricity Demand / Average Heat Demand / Daily H₂ Delivery*		
Minneapolis	6	6.6 ¢/kWh 3.5 ¢/kWh 1.9		558 kW$_e$ 717 kW$_{th}$ 228 kg	217 kW$_e$ 479 kW$_{th}$ 80 kg	515 kW$_e$ 272 kW$_{th}$ 150 kg

Note: Electricity and natural gas prices are commercial rates.

* Fuel cell size set equal to average building electricity demand.

heat production at all times (after heat is subtracted to satisfy the desired level of hydrogen production). This constant electricity and heat demand would persist over the course of the day and night as well as over the course of the year. The result would be maximum total energy efficiency (i.e., minimum total energy use) for the CHHP system and zero use of grid electricity:

$$\text{Total CHHP efficiency} = \frac{\text{Electricity output} + \text{Useful heat output} + \text{Hydrogen energy output}}{\text{Natural gas energy input}}$$

Because of CHHP's higher efficiency compared with conventional electricity sources, and because natural gas is less expensive than grid electricity per unit of energy, this ideal scenario would minimize total energy cost (and thus hydrogen cost compared with the baseline system). It is also likely that the CHHP system's high efficiency and use of natural gas, which is a relatively low-GHG fuel, would reduce GHG emissions relative to the baseline system.

Of course, no such ideal application can exist in the real world. Every geographic location has weather that fluctuates over the course of the day and night and over the course of the year, and every building application has a unique pattern of fluctuating electrical and heat demands. However, considering a hypothetical ideal application helps explain the variations in CHHP performance, cost, and GHG emissions predicted for real-world CHHP applications. In summary, the following optimal operational states for a CHHP system would maximize the energy efficiency of that CHHP system:

- Consistent building electricity demand requiring the appropriately sized fuel cell to run at 100% capacity at all times
- All building electricity demand met by the fuel cell
- Consistent building heat demand consuming all of the fuel cell's useful heat at all times (after heat needed for hydrogen production is subtracted)
- All building heat demand met by the fuel cell

The effect of real-world conditions of climate and use of buildings—and the resulting effects on CHHP energy use and GHG emissions—are explored in the remainder of this chapter, along with variables that affect CHHP hydrogen cost.

7.4 Conditions That Affect CHHP Energy Use and Emissions

This section explores trends related to conditions that affect CHHP energy use and GHG emissions. Geographic location (including climate and the GHG intensity of local grid electricity sources) and building type help determine the viability and benefits of CHHP systems for providing hydrogen fuel and powering buildings. It is also important to match the size of the fuel cell system to building demands. In this section, we present quantitative results to illustrate trends related to these conditions rather than to predict actual cost and performance results. Results are presented in comparison to the baseline system with conventional building energy sources and SMR hydrogen production (described in Section 7.2).

7.4.1 Geographic Location: Climate and Grid Electricity Mix

Climate and the GHG intensity of the local electricity grid mix are two major geographic factors that affect the energy use and GHG emissions of CHHP systems. Climate effects are largely seen in relation to the energy used for building climate control. In these analyses, electrically powered air conditioning provides space cooling, and thermal systems provide space heating and hot water. Buildings in climates with cold winters have greater summer to winter variation in both electricity and heat demand than buildings in milder climates. Winter heat demand is high compared to summer demand, when heat is only needed for water heating. Even in generally colder climates, air conditioning is needed in the summer but is generally not needed in the winter. In contrast, in warm climates, heat is used primarily for hot water year-round, and air conditioning is needed a greater percentage of the time resulting in more consistent loads for both heat and electricity. In the CHHP systems analyzed here, excess heat from the fuel cell is primarily used for hydrogen production. In a conventional CHP application, additional heat would be available for thermally activated chilling (refer back to Figure 7.3).

Figure 7.10 and Figure 7.11 show the influence of climate on hospital CHHP systems in Chicago and Miami. In Chicago, the hospital has significant seasonal demand fluctuations, with large heat demands in the early hours of the winter day and high electricity loads for air conditioning during the summer day. In contrast, the Miami hospital heat demands are relatively flat during the day, and electricity demand is higher in the summer due to air conditioning. The fuel cell in each case follows the electricity load during a given day. During the cold Chicago winter, the electricity demand is well below the fuel cell maximum capacity, causing the fuel cell to be turned down. The low fuel cell electrical output also causes the heat output to be reduced, and the fuel cell meets less than 10% of the heat load. During the Chicago summer, the fuel cell operates at near peak electricity output, providing sufficient electricity except during peak demand in the middle of the day, and heat output supplies nearly half of the heat demand. The patterns of electricity demands and outputs for the mild Miami climate are similar to results for the Chicago climate, with electricity output from the fuel cell being sufficient all day in the winter but not meeting peak demand in the summer. However, the fuel cell's electricity output is more consistently high year-round in Miami. In addition, the fuel cell's heat

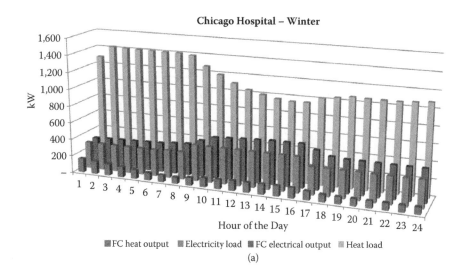

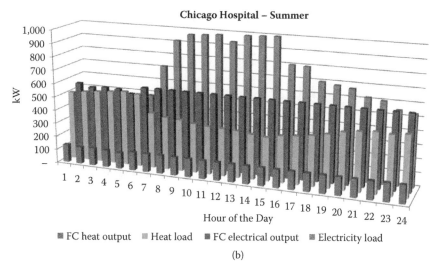

FIGURE 7.10
Chicago hospital loads and fuel cell outputs for a day in winter (A) and summer (B).

output meets a greater fraction of the heat demand in winter and summer in Miami.

The result of these demand profiles tends to be higher capacity of hydrogen production in warmer climates where electricity demand is high due to air conditioning. In these cases, hydrogen production can be increased (on average) as excess heat is used to produce hydrogen. Table 7.4 shows fuel cell sizes and hydrogen production capacities for hospitals in these cities. In each case, the fuel cell size is set to the average electricity demand of the hospital,

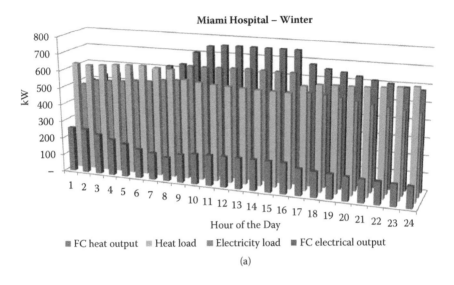

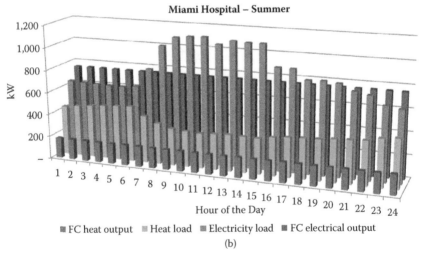

FIGURE 7.11
Miami hospital loads and fuel cell outputs for a day in winter (A) and summer (B).

TABLE 7.4

Fuel Cell Sizes and Hydrogen Production as a Function of Hospital Location

Hospital Location	Fuel Cell Size (kW)	Hydrogen Delivered (kg/day)	Kg H₂/day per kW Fuel Cell Size
Chicago	565	230	0.41
Los Angeles	585	250	0.43
Miami	752	333	0.44

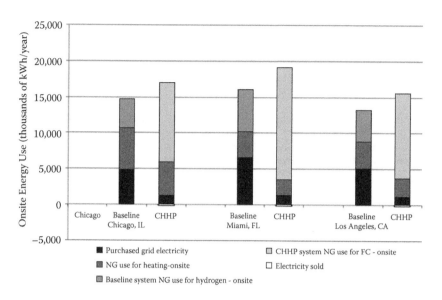

FIGURE 7.12
Onsite CHHP and baseline system energy use for hospitals in Chicago, Miami, and Los Angeles (fuel cell sized to average building electricity demand).

based on previous analyses conducted with multiple runs to identify optimal system sizes (Steward 2009).

Figure 7.12 shows on-site energy use for hospitals with CHHP and baseline systems in Chicago, Miami, and Los Angeles.* The Chicago hospital has the highest building energy use (excluding production of hydrogen) primarily due to high natural gas use for heating in the winter. Los Angeles has the lowest overall building energy use due to lower requirements for heating and cooling in its moderate climate. Chicago and Los Angeles have nearly the same electricity use. Miami has the highest fuel cell utilization and the highest hydrogen production efficiency resulting in the highest rate of hydrogen production (refer back to Table 7.4). Table 7.5 summarizes the CHHP performance for the Chicago, Miami, and Los Angeles hospitals.

Climate interacts with grid GHG intensity to affect the GHG emissions of CHHP and baseline systems in particular geographic locations. Figure 7.13 shows on-site and upstream GHG emissions for CHHP and baseline hospitals in Chicago, Miami, and Los Angeles. Net increases or decreases in upstream and onsite emissions for the CHHP system versus the baseline system are primarily a function of differences in the energy efficiency for hydrogen production between the two systems and substitution of natural gas for electricity in the CHHP system.

---

* Note that on-site energy use is always higher for the CHHP system than for the baseline system because of on-site electricity generation and lower efficiency of hydrogen production.

TABLE 7.5

CHHP Performance for Hospitals in Different Climates (Fuel Cell Sized to Average Building Electricity Demand)

Location	Fuel Cell Utilization	Fuel Cell Efficiency	Hydrogen Production Efficiency*	Fuel Cell Electricity Production Efficiency*	% Heat from Fuel Cell	% of Building Electricity from Fuel Cell	Electricity Sold as Percentage of Total Fuel Cell Electrical Output
Chicago	91.7%	68.2%	25.3%	33.3%	18.2%	74.2%	0.2%
Miami	97.4%	68.5%	25.8%	33.0%	41.3%	78.6%	0.0%
L.A.	94.8%	68.4%	25.6%	33.2%	30.8%	76.7%	0.1%

* Based on breakdown of energy output.

Overall, the higher efficiency of CHHP systems and their replacing of grid electricity with natural gas reduce the GHG emissions of CHHP systems compared with baseline systems. In general, CHHP systems use less total energy than baseline systems because fuel cell electricity generation is more efficient than central power-plant generation, even when the fuel

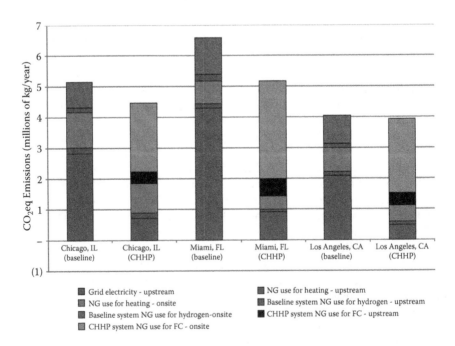

FIGURE 7.13

GHG emissions from hospital CHHP and baseline systems in three cities with different climates and grid electricity GHG intensities (fuel cell sized to average building electricity demand).

cell's electrical efficiency is reduced by producing hydrogen. For the build-ings in this analysis, the fuel cell electrical efficiency was about 41%, or 33% for supply to the building after accounting for electrical use for hydrogen purification and compression (Table 7.5) versus about 31% for the national average grid mix.* On-site use of the available heat also decreases the overall energy use for the building. In addition, CHHP systems replace grid electric-ity with natural gas compared with baseline systems. This reduces CHHP GHG emissions because, in the cities analyzed, the GHG intensity of natural gas is lower than the GHG intensity of the generation mix relied upon for grid electricity (see Section 7.2.3).

Natural gas use is always higher for the CHHP system than for the base-line system, because natural gas is substituted for electricity in the CHHP system and because the baseline system is more efficient than the CHHP system for production of hydrogen. The baseline system's more efficient pro-duction of hydrogen partially offsets the GHG emissions benefits of CHHP. Emissions from natural gas used exclusively for heat production decrease slightly between the baseline system and the CHHP system due to the small amount of waste heat from the CHHP system that is used for building demand. However, overall emissions from natural gas use increase when moving from the baseline system to the CHHP system.

Figure 7.14 illustrates the net GHG emissions of hospital CHHP systems versus baseline systems in the three cities. Negative values indicate lower emissions for the CHHP system compared with the baseline system. Positive values indicate emissions increases for the CHHP system. The overall net emissions, in metric tonnes CO_2eq/year, are shown above the bars. The base-line system has lower efficiency associated with electricity production, and higher total emissions, than the CHHP system. In addition to the efficiency differential between the CHHP system and the baseline, the magnitude of emissions reduction for the CHHP system depends on the emissions associ-ated with the competing grid mix. Of the three cities, Miami has the high-est GHG emissions factor (refer back to Table 7.2). It is almost 60% higher than the GHG emissions factor for the California grid, which has a large proportion of hydropower and renewable generation. As the Los Angeles case demonstrates, distributed CHHP is not always clearly preferable to conventional electricity/heat supply and SMR hydrogen production from a GHG-reduction perspective.

Analysis presented in Figure 7.15 isolates the effect of grid GHG intensity. It shows the sensitivity of CHHP's GHG emissions advantage to hypothetical changes in the grid GHG intensity for a hospital in Chicago. The bar on the far left shows the net GHG emissions advantage of CHHP versus the baseline system with the current/real Chicago grid mix. Each bar to the right shows the net GHG emissions advantage as the GHG intensity of the grid mix is reduced by 5%. For each 5% reduction in grid GHG intensity, the CHHP's

---

* http://www.epa.gov/chp/basic/efficiency.html.

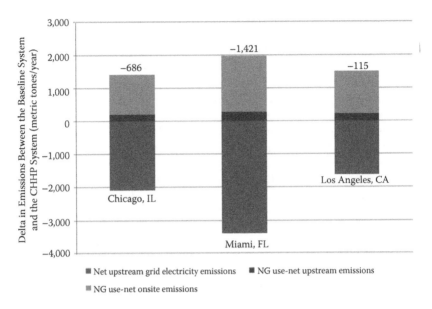

FIGURE 7.14
Net GHG emissions of CHHP systems vs. baseline systems for hospitals in three cities with different climates and grid electricity GHG intensities (fuel cell sized to average building electricity demand).

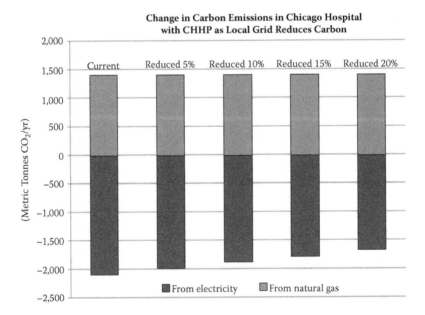

FIGURE 7.15
Net GHG emissions change from a hospital CHHP system vs. a baseline system as the GHG intensity of the grid electricity mix is reduced from the current Chicago mix to mixes with 5%, 10%, 15%, and 20% lower GHG intensity.

electricity emissions savings decrease about 5%. Because the natural gas emissions remain constant, the net GHG emissions advantage of the CHHP system decreases by 15%, 18%, 22%, and 28% as the assumed electricity grid GHG intensity decreases by each 5% increment.

7.4.2 Building Type

Like climate, building type influences electricity and heat demand, which affects the performance of CHHP systems. In general, CHHP systems in buildings with the most constant energy demand throughout the day and throughout the year are the most energy efficient and yield the largest GHG emissions savings versus a baseline system. Three building types—hospitals, large offices, and large hotels—were selected to illustrate the effect of building use profiles on the performance of CHHP systems.

Figure 7.16 shows the energy use resulting from the CHHP and baseline systems for three building types in Chicago and Miami, with the fuel cell sized to average building electricity demand. In all cases, the on-site energy use for the CHHP system is higher than for the baseline system because of

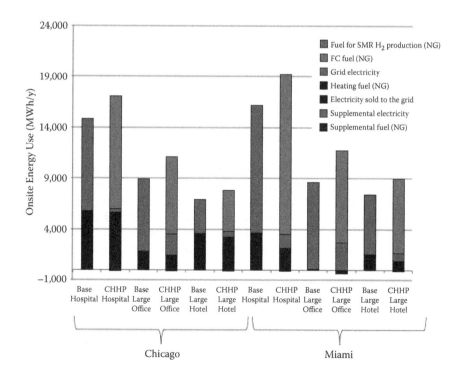

FIGURE 7.16
CHHP and baseline system onsite energy use for different building types in Chicago and Miami (fuel cell sized to average building electricity demand).

on-site electricity generation and lower hydrogen production fuel efficiency for the CHHP systems. Of the three building types, hospitals, which have 24-hours-per-day occupancy and activity levels, have the smallest hourly fluctuations in both electricity and heat demand. Large offices, which are occupied only during the day, have the largest fluctuations in electricity and heat demand, and hotels, with 24-hour occupancy but little activity during the night, fall in between. The standard deviation in electricity demand as a percentage of the average can be used to quantify the magnitude of demand fluctuations (Table 7.6). The primary effect of large hourly fluctuations in electricity demand is reduced utilization of the fuel cell. For hospitals in Chicago and Miami, fuel cell utilization is 92% and 97%, respectively. Fuel cell utilization decreases to 71% and 68% for offices in the same cities. Electricity is also sold to the grid for offices in both cities. This occurs because the electricity demand has occasionally fallen below the minimum turndown level for the fuel cell, and electricity generated during this time must be fed back onto the grid. This situation may not be economically favorable because fuel must be consumed to generate electricity when electricity demand generally is low and electricity must be sold for a low price. Hotels and hospitals have high heat demand relative to electricity demand because of high demand for hot water. These building uses are favorable for CHP systems because a greater percentage of the fuel cell's heat output can be used to supply this demand. Heating of water is also somewhat decoupled from hot water demand so waste heat from the fuel cell can be used even when hot water demand is low. Nearly all of the fuel cell's heat output can be used for hotels and hospitals in both cities (Table 7.6). In contrast, fuel cell heat utilization for offices is much lower. For the office building in Miami, only 9% of the usable heat from the fuel cell CHHP system can be used to meet the building demand. Low heat utilization reduces the total energy efficiency for the CHHP installation.

Figure 7.17 shows CHHP emissions savings versus the baseline system by building type in Chicago. The emissions savings are greatest for the hospital where the fuel cell is most effectively used to meet both the building electricity and heat demand. The large office and hotel have similar emissions savings relative to the baseline system. In the office, electricity makes up a larger fraction of the total on-site energy use than in the hotel. Because conventional electricity generation has high upstream emissions relative to the emissions from onsite generation using a fuel cell, the office application benefits more from installation of the fuel cell, which compensates for lower utilization and overall efficiency. Although more of the heat from the fuel cell is utilized in the hotel (100% see Table 7.6) than in the large office application (53%), almost the same amount of heat energy is supplied by the fuel cell in the two buildings and the fuel cell in the office supplies a higher percentage of the building heat demand.

TABLE 7.6

Load Characteristics and Fuel Cell Performance for Buildings in Chicago and Miami

Building	Total Electricity Demand (MWh/y)	Avg Electricity Demand (kW)	One STDev as% of Avg Electricity Demand	Avg Heat Demand (kW)	Heat Provided by Fuel Cell (% of Demand)	Fuel Cell Heat Utilization	Fuel Cell Utilization*	Fuel Cell Total Efficiency
Chicago								
Hospital	4,948	565	36%	660	18.2%	100%	92%	68%
Large Office	4,532	517	78%	212	20.2%	53%	71%	63%
Large Hotel	1,933	221	49%	415	10.5%	100%	86%	68%
Miami								
Hospital	6,588	752	25%	419	41.3%	100%	97%	69%
Large Office	5,547	633	79%	13	69.3%	9%	68%	59%
Large Hotel	3,185	364	31%	180	43.3%	98%	94%	68%

* Electrical output/output at rated power.

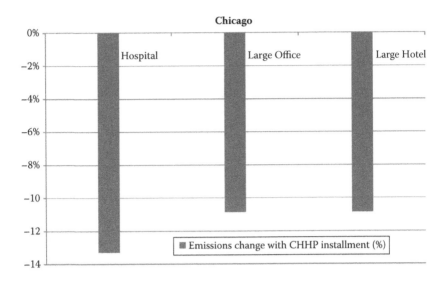

FIGURE 7.17
CHHP emissions savings vs. baseline system by building type, Chicago (fuel cell sized to average building electricity demand).

7.4.3 Size of Fuel Cell System Relative to Building Electricity Demand

In addition to geographic location and building type, the size of the CHHP system relative to building electricity demand affects the resulting performance and GHG emissions. To illustrate this, we modeled a hospital in Chicago with a CHHP fuel cell size (output in kW of electricity) set equal to the average building electricity load minus ½ standard deviation, the average load, the average load plus ½ standard deviation, and the average load plus 1 standard deviation. Figure 7.18 shows the results.

Fuel cells that are large relative to the building's electricity demand produce more hydrogen and meet more of the building's demand for electricity and heat, but they are used at full capacity less often than smaller fuel cells. As the size of the fuel cell increases, the size of the comparison baseline SMR system also increases, which increases its energy use and GHG emissions.

All the electricity produced by the fuel cell displaces grid electricity. Because the fuel cell is electricity load following, if the electricity load is always high in relationship to the fuel cell size, all or nearly all of the electricity produced by the fuel cell will displace electricity that otherwise would have been purchased for the building. A small amount of "excess" electricity may be generated if the fuel cell cannot turn down production fast enough to match a rapid change in the building load. "Excess" electricity will also be generated if the building electricity demand falls below the minimum production rate of the fuel cell (20% of its rated power in this case).

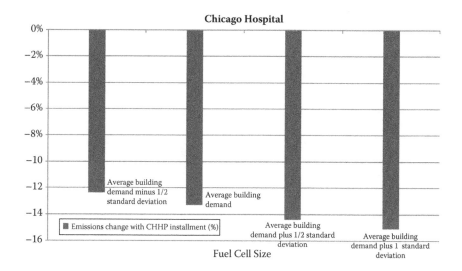

FIGURE 7.18
CHHP emissions savings vs. baseline system by relative fuel cell size, Chicago hospital.

Hospitals have a relatively flat 24-hour electricity load profile, consuming nearly as much electricity during the night as during the day. Even large fuel cells in hospitals do not ramp up and down significantly and are never turned down to their minimum level. Therefore, little if any excess electricity is generated. Figure 7.18 shows that GHG emissions savings of the CHHP versus the baseline system increase with increasing fuel cell size. The effects of relative fuel cell size vary by building type and climate (see Figure 7.22 in the summary section for examples).

7.5 Factors That Affect CHHP Hydrogen Cost

Many factors affect the economics of CHHP systems but not their performance, such as financing methods, expected rate of return, inflation rate, and government incentives. We do not explore the effects of varying these factors in this chapter. Instead, we focus on the effects on hydrogen cost of CHHP system performance—as discussed in Section 7.4—as well as regional electricity and natural gas prices, which are the external factors with the largest effect on CHHP hydrogen cost.

7.5.1 CHHP System Performance

Figure 7.19 shows the effect of climate on hydrogen cost from hospital CHHP systems in various cities, compared with baseline systems. To isolate the

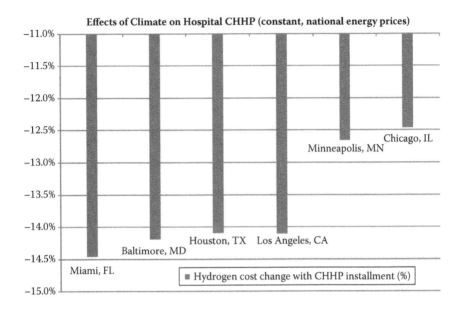

FIGURE 7.19
Difference between CHHP cost of hydrogen vs. baseline system for hospitals in various climates, fuel cell sized to average building electricity demand (natural gas at \$3.60/mmBtu and electricity at 11.5¢/kWh for each city to isolate climate effects).

effect of climate, we artificially set energy prices in each city to be equal at average national values. As expected, the general trend is for the differential benefit of hydrogen to decrease as the climates become more variable between summer and winter as indicated by the ratio of one standard deviation to the total electricity demand (Table 7.7). For more variable electrical loads, the fuel cell utilization and the average amount of hydrogen produced per kW of fuel cell rated power decrease. Both trends increase the cost of hydrogen.

Figure 7.20 shows how the cost of CHHP-produced hydrogen increases with increasing fuel cell size. Full utilization of a fuel cell system tends to result in the most favorable economics, and the fuel cells sized at and below average electricity demand produce the lowest-cost hydrogen in this Chicago hospital. As the fuel cell size is increased, the capital costs of the CHHP system increase and the fuel cell is underutilized, resulting in higher-cost hydrogen (and a higher volume of hydrogen produced). The effects of relative fuel cell size vary by building type and climate. For very small installations, the cost for storing and dispensing hydrogen, which are relatively fixed for a range of system sizes, are a significant fraction of the cost of the installation. For these systems, increasing the fuel cell system size relative to the building load decreases overall costs per unit of energy output due to economies of scale that compensate for the decreased system utilization.

TABLE 7.7

Hydrogen Production and Fuel Cell Sizes for Hospitals in Various Cities Ranked by Fuel Cell Size (Fuel Cell Sized at Average Electrical Load)

Location	Fuel Cell Size (kW)	Fuel Cell Utilization (Electrical Output/ Output at Rated Power)	Hydrogen Output (kg/day)	Hydrogen Output per kW Fuel Cell (kg/ day-kW)	One STDev as% of Avg Electricity Demand
Miami, FL	752	97.4%	333	0.44	25%
Houston, TX	714	96.0%	310	0.43	27%
Baltimore, MD	631	95.2%	271	0.43	30%
Los Angeles, CA	585	94.8%	250	0.43	31%
Chicago, IL	565	91.7%	230	0.41	36%
Minneapolis, MN	558	91.9%	228	0.41	36%

Compare the cost results in Figure 7.20 with the GHG emissions results in Figure 7.18, which shows the GHG emissions savings of CHHP versus the baseline system increasing with increasing fuel cell size. This illustrates the trade-off between cost and GHG emissions that must be balanced in some CHHP systems. These trends suggest that future market conditions that include a carbon tax would favor larger fuel cell systems, especially if carbon credits are also attained from transportation fuels markets.

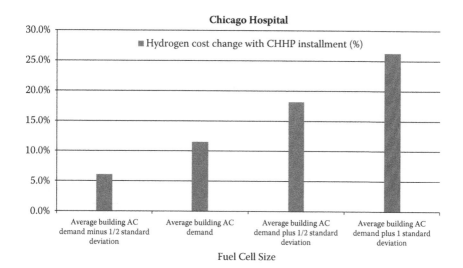

FIGURE 7.20

Increased cost of CHHP-produced hydrogen vs. baseline system-produced hydrogen by relative fuel cell size, Chicago (local electricity and natural gas prices).

7.5.2 Regional Electricity and Natural Gas Prices

Regional variations in electricity and natural gas prices impact the economics both directly and indirectly. Natural gas prices directly impact the economics of both the CHHP system and the baseline system. Higher natural gas prices result in higher costs for both systems because natural gas is used as a feedstock for SMR and as a heating fuel and fuel for the fuel cell. Electricity prices also have a direct impact on the baseline system and the supplementary electricity that must be purchased for the CHHP system. However, the benefit derived from producing electricity from the fuel cell rather than buying it from the grid increases with increasing grid electricity price, making the fuel cell system more economically favorable in comparison to the baseline system. The ratio of the price of electricity to natural gas provides an estimate of the economic benefit (refer back to Figure 7.9). A higher ratio indicates more favorable economics. Figure 7.21 illustrates the sensitivity of hydrogen cost to natural gas and grid electricity prices for several buildings in Chicago. Increasing the natural gas price increases the cost of hydrogen through increasing the fuel cost. In contrast, increasing the grid electricity price decreases the cost of hydrogen because the electricity produced by the fuel cell becomes more valuable as it displaces more expensive grid electricity.

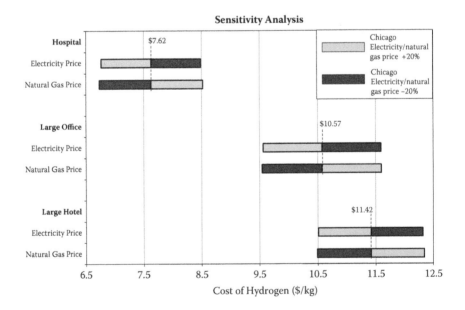

FIGURE 7.21
Sensitivity of hydrogen cost to electricity and natural gas prices for three building types in Chicago (fuel cell sized to average building electricity demand); central values are at actual Chicago electricity and natural gas prices, end values are at +/–20% of actual prices.

7.6 Summary of Trends Affecting CHHP Performance and Cost

Fuel cell–based CHP installations with a side-product of hydrogen were explored in this chapter. The effects of climate and building type and use on building loads and consequently on the performance of the CHHP systems were compared to a baseline system in which electricity and heat are provided by conventional means and hydrogen is produced through on-site SMR. External factors that determine the benefit derived from installation of a CHHP system in a given building relative to the baseline system were also analyzed. These factors include natural gas and electricity prices as well as the GHG emissions associated with the regional grid generation mix.

Climate affects overall building electricity and heat loads and the timing and variability of these loads with respect to each other. In general, milder climates result in more uniform electricity and heat loads than more variable cold-winter climates. More uniform building loads also improve the energy efficiency and utilization of the CHHP system, resulting in lower overall energy use and lower costs.

Building type and use affects building loads similarly to climate. Building types, such as hospitals, that have high occupancy and activity throughout the day and night tend to have lower delivered energy cost for CHHP installations than buildings with more variable occupancy such as office buildings that are only occupied during the day.

Installing a CHHP system that is large relative to the building average electricity load tends to increase costs because the fuel cell is required to ramp up and down in output, thus decreasing utilization. This effect is more pronounced for climates and building uses that produce large variations in load between day and night and seasonally.

The magnitude of the benefit of installing a CHHP system relative to a baseline system supplying the same energy products is largely determined by external factors, including fuel and competing electricity prices and GHG emissions associated with the regional electricity generation mix. High local grid electricity prices make the electricity product of the CHHP system more valuable and improve the economics of the system relative to the baseline. High natural gas prices increase costs for both the baseline system and the CHHP system, but the CHHP system economics are more sensitive to natural gas price because more natural gas is used in the CHHP system per unit output of energy. In the CHHP system, natural gas is used to produce electricity as well as heat and hydrogen. In addition, hydrogen production is more efficient using SMR than it is in the fuel cell system, so less natural gas is used in the baseline system. Total GHG emissions are the sum of on-site emissions associated with combustion/reforming of natural gas and upstream emissions associated with generation of electricity and, to a much lesser extent, production of the natural gas. As discussed previously, on-site energy use is always higher for the CHHP system than for the baseline system (see Figure 7.12 and Figure 7.16). Therefore, the GHG emissions benefit (or penalty) for installation of the CHHP system can be expressed as the ratio between the emissions savings from substituting fuel cell–generated electricity for grid electricity and the emissions penalty for increased natural gas use:

$$\frac{\{(\text{baseline grid electricity use} - \text{net supplemental grid electricity use for CHHP}) \times \text{Electricity GHG emissions factor}\}}{\{(\text{CHHP natural gas use} - \text{baseline natural gas use}) \times \text{natural gas GHG emissions factor}\}}$$

A ratio greater than one indicates a net savings in GHG emissions, and a ratio less than one indicates a net increase in GHG emissions.

Figure 7.22 illustrates the combined effects of climate, building use, fuel cell system size, and local natural gas and electricity prices on the levelized cost of hydrogen produced from these installations. Many competing factors influence the economics for each installation. In general, applications with larger and more uniform average electricity loads result in better economic performance. High local electricity prices and low natural

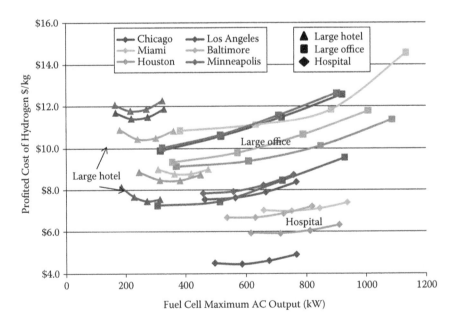

FIGURE 7.22
Profited cost of hydrogen by location, building type, and fuel cell system capacity.

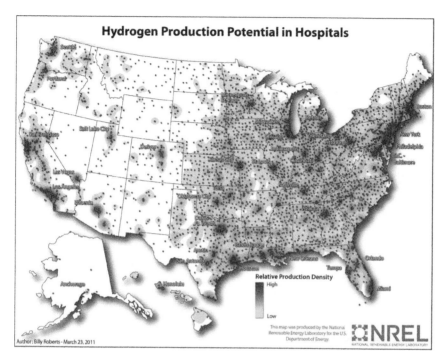

FIGURE 7.23
Hospitals and relative hydrogen production potential from CHHP.

gas prices also improve the economics for CHHP systems relative to the baseline. The low costs for all buildings in Los Angeles relative to other locations are a good example of the effect of "spark spread" on economics for these systems.

The potential for small-scale hydrogen production from CHHP installations in hospitals in the United States is illustrated in Figure 7.23. One of the primary advantages of this strategy is that hydrogen would be available at numerous, probably convenient, locations in metropolitan areas, albeit in small quantities. Because hospitals are also well distributed in more rural locations, CHHP installations at hospitals would also provide refueling between major population centers. The number of fuel cell vehicles that could be supported using this strategy in select metropolitan areas is shown in Table 7.8, and this support would prove significant during the early introduction of vehicles. Greenhouse gas reductions would also be a benefit of these installations. Each building has four data points related to the building's average electricity demand, from left to right: −1/2 standard deviation of average electricity demand, average electricity demand, +1/2 standard deviation of average electricity demand, +1 standard deviation of average electricity demand.

TABLE 7.8

Potential Hydrogen Production, GHG Emissions Savings, and FCEVs Supported for Hospital CHHP Installations in Select U.S. Metropolitan Areas

Metro Area	Number of Hospitals	H$_2$ Production (Metric Tonnes H$_2$/year)	Emissions Savings (Metric Tonnes CO$_2$/yr)	FCEVs Supported
Miami	82	9,967	116,516	49,833
Houston	117	13,239	177,761	66,193
Los Angeles	163	14,874	18,820	74,369
Baltimore	33	3,264	49,484	16,321
Chicago	137	11,501	93,986	57,506
Minneapolis	41	3,412	57,599	17,060
Fairbanks	2	147	1,586	737
Denver	33	2,770	22,639	13,852
Portland	27	2,671	40,487	13,354
San Francisco	65	5,931	7,505	29,656

7.7 Conclusions

High-temperature fuel cells in CHP applications can be configured to produce excess hydrogen for use in vehicles. In producing three energy products, these tri-generation systems are referred to here as CHHP systems. We have modeled the energy, GHG, and hydrogen cost implications of this technological strategy for facilitating efforts to establish a fueling infrastructure to support early hydrogen vehicle markets. Our analysis employs the FCPower model, which was developed by the National Renewable Energy Laboratory and is available for download as an Excel spreadsheet. This chapter explains some of the basic modeling assumptions underlying the representation of MCFC systems in the FCPower model and reviews the total energy use, emissions, and hydrogen cost for CHHP installations in comparison to conventional supplies of energy to buildings and small-scale dedicated (SMR) production of hydrogen.

Through these analyses, we explore the effects of climate, geographic location, grid carbon intensity, building type, and fuel cell size on CHHP efficacy. In general, the building electricity load dictates the behavior of the CHHP system. Consistent electricity demands such as those in milder, warm climates better utilize fuel cell capacity and are able to generate more excess heat and hydrogen with reduced emissions. In climates that fluctuate from season to season, fuel cell capacity is only fully utilized in the summer, and there is less excess heat available for the building and hydrogen generation in the winter, resulting in a lower energy and GHG emissions reduction advantage. In addition, assumptions regarding local grid GHG intensity affect the apparent advantage of CHHP systems. As the GHG intensity of the displaced electricity supply increases, the GHG emission reductions achieved by CHHP systems also increase. Taking into account all three energy products from a CHHP system, we generally find GHG emission reductions of approximately 10 to 20% for installations in which the fuel cells were sized at the building average electrical load. However, the emissions savings are very dependent on the regional grid generation mix. For buildings in Los Angeles, where the regional grid mix is very "green," emissions reductions were negligible.

In terms of building types, buildings like hospitals or hotels with more consistent electricity loads and fuel cell utilizations see greater energy and emissions advantages than buildings with fluctuating loads, like office buildings and schools. The analysis presented here was necessarily limited to only a few building types. However, stationary fuel cells could provide benefits in a wide variety of applications. For example, fuel cell CHHP systems could be used for supplying hydrogen for home refueling and CHP for larger residential applications, especially where district heating or providing heat for a community pool or recreation center might also be included. Facilities with large cooling loads, such as data centers or supermarkets, might also be good applications for CHHP.

Previous studies have identified on-site SMR as a key least-cost pathway in many infrastructure rollout scenarios. In the support of near-term hydrogen vehicle markets, many stations will likely be relatively small, supplying between 100 and 700 kg of hydrogen per day. At these small volumes, SMR technologies are somewhat more costly due to economies of scale and potential low utilization factors during periods of low demand; therefore, CHHP systems are more likely to prove competitive. As the market for hydrogen vehicles expands, larger refueling stations would be required, with on-site SMR or other production and delivery technologies achieving larger economies of scale. However, the financial and technological risks associated with small, early stations are a significant barrier to the deployment of hydrogen vehicles, so identification of lower risk and lower capital pathways is an important area for analysis and technology development. By examining the various factors that affect hydrogen cost for CHHP systems, we are able to propose general rules of thumb for situations in which hydrogen from CHHP systems may prove competitive with onsite SMR stations (Table 7.9).

A down arrow ($\downarrow$) means the value of the characteristic (hydrogen price, life cycle energy use, or life cycle GHG emissions) typically decreases for the CHHP system in relation to the baseline system under the specified condition. An up arrow ($\uparrow$) means the value of the characteristic typically increases for the CHHP system in relation to the baseline system under the specified condition.

TABLE 7.9

Typical Effect of Conditions on Electrical-Load-Following CHHP System vs. Baseline System

Condition	Hydrogen Price	Life Cycle Energy Use	Life Cycle GHG Emissions
Climate more stable (less seasonal variation) and warm	↓	↓	↓
More carbon-intense grid electricity sources	N/A	N/A	↓
Building electricity demand more constant	↓	↓	↓
Larger fuel cell relative to building electricity demand (lower fuel cell utilization)	↑	↓	↓
Electricity price higher	↓	N/A	N/A
Natural gas price lower	↓	N/A	N/A

References

DOE. 2009. *Delivering Renewable Hydrogen: A Focus on Near-Term Applications*, Palm Springs, CA, U.S. Department of Energy, National Renewable Energy Laboratory, California Fuel Cell Partnership, Retrieved 9-10-2010, from http://www1.eere.energy.gov/hydrogenandfuelcells/delivering_hydrogen_wkshp.html.

DOE. 2011. DOE Fuel Cell Power Analysis (http://www.hydrogen.energy.gov/fc_power_analysis.html), DOE H2A Analysis (http://www.hydrogen.energy.gov/h2a_analysis.html), accessed March 2011.

EEA. 2007. CHP Installation Database. Energy and Environmental Analysis Inc., developed for Oak Ridge National Laboratory.

EPA. 2010. "eGRID." Retrieved 9-10-2010, from http://www.epa.gov/cleanenergy/energy-resources/egrid/index.html.

EG&G. 2005. *Fuel Cell Handbook*, 7th edition, EG&G Technical Services, Inc., University Press of the Pacific.

EPRI and Southern Company Services. 2003. *Technical and Economic Assessment of Combined Heat and Power Technologies for Commercial Customer Applications*. EPRI: Palo Alto, CA; Southern Company Services: Birmingham, AL.

EPRI. 2004. *Update on Fuel Cell Development; Review of Major and Stealth Fuel Cell Players' Activities: Stealth Player Reviews*. EPRI: Palo Alto, CA.

EPRI. 2005. *Technology Review and Assessment of Distributed Energy Resources: 2005 Benchmarking Study*. EPRI: Palo Alto, CA.

Field, K., M. Deru, and D. Studer. 2010. *Using DOE Commercial Reference Buildings for Simulation Studies*, Preprint. SimBuild 2010. New York, New York, National Renewable Energy Laboratory.

Greene, D. L., P. N. Leiby, et al. 2008. *Analysis of the Transition to Hydrogen Fuel Cell Vehicles & the Potential Hydrogen Energy Infrastructure Requirements*. Oak Ridge National Laboratory. Report number: ORNL/TM-2008/30.

ICF International. 2008. *Availability, Economics, and Production Potential of North American Unconventional Natural Gas Supplies*.

IPHE. 2010. IPHE Infrastructure Workshop: Workshop Proceedings, Sacramento, CA, February 25–26, 2010, International Partnership for the Hydrogen Economy, California Fuel Cell Partnership, National Renewable Energy Laboratory. Retrieved 9-10-2010, from http://www.iphe.net/docs/Events/iphe_infrastructure_workshop_feb2010.pdf.

Melaina, M. W. 2003. "Initiating hydrogen infrastructures: preliminary analysis of a sufficient number of initial hydrogen stations in the U.S." *International Journal of Hydrogen Energy* 28(7): 743–755.

Melaina, M. W., and J. Bremson. 2006. *Regularities in Early Hydrogen Station Size Distributions. Energy in a World of Changing Costs and Technologies*, 26th North American Conference, International Association of Energy Economics, Ann Arbor, Michigan.

NRC. 2008. *Transitions to Alternative Transportation Technologies: A Focus on Hydrogen*. National Academies Press. National Research Council of the National Academies, Committee on Assessment of Resource Needs for Fuel Cell and Hydrogen Technologies, Washington, D.C.

Patel, P., and T. Leo (Fuel Cell Energy). 2009. Personal communication. July 16, 2009.

Rashidi, R. et al. 2009. Performance investigation of a combined MCFC system. *International Journal of Hydrogen Energy*, 34(10): pp. 4395–4405.

Shipley, A., A. Hampson, et al. 2008. *Combined Heat and Power: Effective Solutions for a Sustainable Energy Future*. Oak Ridge National Laboratory. Report number: ONRL/TM-2008/224, Retrieved 9-10-2010, from http://apps.ornl.gov/~pts/prod/pubs/ldoc13655_chp_report____final_web_optimized_11_25_08.pdf.

Steward, D. 2009. *The Influence of Building Location on Combined Heat and Power/Hydrogen (Tri-Generation) System Cost, Hydrogen Output and Efficiency*. National Hydrogen Association Conference and Exposition. Columbia, SC, March 30 to April 3, 2009.

8

Hybrid and Plug-in Hybrid Electric Vehicles

Andrew Meintz

CONTENTS

8.1 Introduction

Hybridization of a vehicle propulsion system can improve performance and efficiency by splitting the power required to propel a vehicle between multiple energy sources. The essential improvement of a hybrid electric vehicle (HEV) is the more efficient use and design of the primary energy source (e.g., internal combustion engine, fuel cell) than can be obtained in a conventional vehicle. In conventional vehicles, the internal combustion engine (ICE) acts as the only source of power. This single source design is less efficient since the ICE must operate at low power or idle speeds where it consumes more fuel or consumes fuel without providing power for propulsion. As a result, only 10 to 15% of the energy contained in gasoline is used to propel the vehicle [1]. In a hybrid propulsion system, the power demand of the primary energy source can be decoupled from the instantaneous road load requirements through the use of a secondary energy source(s) (batteries, electrochemical capacitors, flywheels, etc.) resulting in improved efficiency.

A regenerative secondary source allows for decoupling of the primary source through bidirectional energy flow in which power from the primary source can be greater than or less than the power required by the road load. Additionally, this bidirectional flow can allow for energy to be recaptured from the road when slowing or stopping the vehicle through regenerative braking. Decoupling of the primary source is limited by three characteristics of the secondary source(s):

1. The secondary source must regain any energy used from the primary energy source.
2. Power of the secondary source is constrained based on its interface to the propulsion system.
3. The secondary source must constrain its energy state to retain adequate discharge and recharge power.

A conceptualized hybrid electric vehicle, as shown in Figure 8.1, illustrates how energy can flow between sources and the road [2].

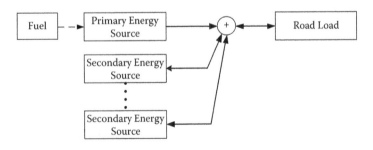

FIGURE 8.1
Conceptualized HEV.

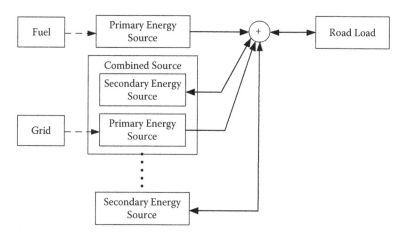

FIGURE 8.2
Conceptualized PHEV.

Hybrid electric vehicles allow for increased performance and efficiency through the combination of a primary energy source with a secondary source(s) of energy. It follows from this concept that multiple primary energy sources may be combined with a secondary source to allow for even greater flexibility in the use of primary energy sources. An example of this concept is a plug-in hybrid electric vehicle (PHEV). The PHEV can utilize the primary energy sources independently or in a combined method with a secondary source(s) to meet the road load requirement. In practice, additional primary source energy is in the form of increased capacity of a secondary source (e.g., battery). A portion of the energy is provided by recharge from the electric grid and not from the primary source during the drive cycle. This use of the primary energy of a battery is known as charge depleting (CD) operation; likewise, the use of secondary energy is known as charge sustaining (CS) operation. In Figure 8.2, a conceptualized plug-in hybrid electrical vehicle with two primary sources and a secondary source illustrates how energy can flow between sources and the road.

8.2 Performance Characteristics

The requirements of a vehicle propulsion system are of importance when examining changes in the composition of energy sources. The vehicle must still meet specific drivability criteria to obtain consumer acceptability such that vehicle acceptance in the market can occur. While there are many criteria that are considered in consumer acceptability, only the following are of consideration when comparing different hybrid vehicle architectures.

8.2.1 Fuel Economy

The fuel economy of a hybrid vehicle is a difficult property to quantify especially when comparing different vehicle fuel technologies. The use of grid electricity as an additional primary energy source for plug-in hybrid vehicles complicates the designation of how many miles per gallon (or L/100 km) a vehicle can obtain.

Determining a combined fuel economy for a vehicle operating in both charge sustaining and charge depleting mode requires a method to assign an equivalent gasoline value for the grid electrical energy (AC kWh) used to charge the battery system. A direct energy comparison of grid electrical energy to the energy content of gasoline does not produce an equitable result, since this assumes that the energy capacity and use of the battery system is sufficient for any duration of travel. This further complicates the matter, as a direct energy comparison does not reflect the differences in cost from the two energy sources.

Additionally, the charge-depleting range that a vehicle can achieve coupled with the distance a consumer will likely drive can influence the real-world fuel economy result. One method to account for these differences is to create a weighted fuel economy based on aggregate U.S. vehicle fleet driving data. In this way a daily driving distance of the average consumer can be used to define how far beyond the depleting range a vehicle is likely to be driven. This allows for the fuel economy of a vehicle in both charge depleting and charge sustaining to be averaged over the total driving distance to create a combined fuel economy [3].

8.2.2 Range

Range of a vehicle is the distance the vehicle can travel using the onboard energy storage. The distance will vary based on vehicle speed and road conditions. The range of the vehicle is an important consideration in vehicle propulsion systems where energy storage refills or recharges present unique time and infrastructure requirements.

8.2.3 Acceleration

Acceleration time is a measure of how long it takes a vehicle to increase its speed and is typically measured from 0 to 60 mph and 50 to 60 mph on level, 3%, and 6% grades. This criterion is related to maximum peak power available at the road from all propulsion system sources for a short period of time.

8.2.4 Gradeability

Gradeability is the maximum percent grade on which a vehicle can travel for a specified time at a specific speed. The gradeability at speed measure is the maximum speed at which the vehicle can maintain on a grade of 3% and 6%

[4]. This criterion is related to maximum power available at the road from all propulsion system sources for an extended period of time. The gradeability limit measure is the calculated grade on which the vehicle can start from a stop and climb [4].

8.3 Hybrid Vehicle Architectures

How energy from primary and secondary sources is combined in the vehicle is of particular interest as it affects the inherent performance and efficiency of the vehicle. This combination can be realized in a fully electrical, fully mechanical, or joint electrical and mechanical method. These combination structures are known as series, parallel, and series-parallel [5]. The use of these hybrid architectures in different vehicle applications has an inherent implication on the optimal use of the primary energy source, the power ratings of the components, and the overall performance. In the following sections, vehicle performance criteria will be explored with respect to the electrical and mechanical power-split architectures.

8.3.1 Electrical Power-Split Architectures

Electrical power-split architectures combine power from multiple electrical sources using power electronic converters and inverters on one or multiple high-voltage buses. The power sources can be the internal combustion engine, the vehicle wheels (regenerative) coupled through an electric machine, or electrochemical such as batteries, capacitors, and fuel cells.

8.3.1.1 Engine Series Hybrid

The engine series hybrid is a vehicle architecture that uses an internal combustion engine tied to a generator as the primary source and a battery as the secondary source as shown in Figure 8.3. This architecture is considered to have an electric power-split because it combines energy on an electric bus and can use power from either the primary source (P_{GE}) or the secondary source (P_B) to drive the traction motor (P_{ME}) and propel the vehicle (e.g. P_{GE} + $P_B \approx P_{ME}$). This combination method allows for two degrees of freedom in the operation of the power flow balance in the vehicle. The balance of power on the electrical node (e.g. $P_{GE} + P_B \approx P_{ME}$) can be altered depending on the requirements of the road load (P_R) with consideration of the battery state of charge (SOC) or energy content and the efficiency of the associated power demand from the generator (P_{GE}). The operation of the combustion engine speed can be varied through the use of the generator inverter and engine control. At a given power requirement (P_{GE}), the speed of the engine can be

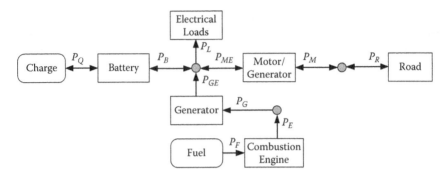

FIGURE 8.3
Engine series hybrid.

controlled such that it operates at the most efficient torque/speed point for the combustion engine.

The advantage of the engine series design is that the combustion engine is able to run at its optimal speed and torque resulting in lower fuel consumption. However, the two energy conversion stages (engine/generator and motor/generator) result in the transformation of all energy between the combustion engine and the road load. This results in better application of this design for consistent low power use such as in city driving where the losses of idle engine use outweigh the transformation losses. [1]

8.3.1.2 Fuel Cell Series Hybrid

Fuel cell series hybrid vehicle architecture uses two electric buses that are linked through a dc-dc converter as seen in Figure 8.4. The dc-dc converter is used to connect the differing voltages of the battery system and the fuel cell system. These differing voltages are the result of two independent electrochemical

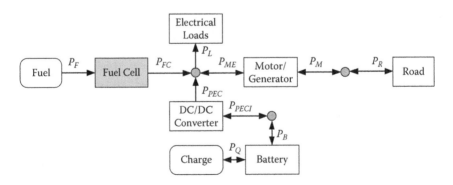

FIGURE 8.4
Fuel cell series hybrid.

processes that vary based on power load from each system. The dc-dc converter allows for a controllable power flow from each system to be combined at a common electrical node. The power flow of this vehicle architecture is similar to the engine series case, where the power from the fuel cell (P_{FC}) and the power from the battery (P_{PEC}) are used by the motor/generator (P_{ME}) to meet the road load requirement (P_R). The power flow balance of the vehicle (e.g. $P_{FC} + P_{PEC} \approx P_{ME}$) can be freely chosen based on the battery SOC and the efficient operation of the fuel cell. In this design the energy from the primary source does not require two energy conversions as the inverter of the motor/generator can be designed to operate within the voltage range of the fuel cell.

8.3.1.3 Electrical Power-Split Performance

The performance of electrical power-split vehicle architectures depends greatly on the power capability of the traction motor/generator. This motor is required to supply all traction power for the vehicle and must be sized to meet the peak power requirements for acceptable acceleration times. Furthermore, the dynamic response of the primary source and the peak power limitations of the secondary source contribute to the combined peak power available to the motor. Secondary energy sources such as electrochemical capacitors can provide significant instantaneous power but are limited by the total amount of energy that can be delivered. For battery systems the amount of instantaneous power is limited by the chemistry, state of charge, and size of the system. In the fuel cell series case the secondary power system is further constrained by the power limitation and transient response of the dc-dc converter.

Gradeability of these architectures is related to the maximum sustained power rating of the traction motor as well as the maximum power of the primary power source. The secondary source is typically too small to meet significant sustained power requirements so the primary source and conversion mechanism must be sized to meet the power requirements of the gradeability criteria. The conversion mechanism, such as the generator/inverter in the engine series hybrid, must be designed to meet this sustained power requirement. These conditions can be relaxed if an additional primary source, such as in a PHEV, is available to provide sustained power. Range is affected by the amount of fuel storage available on the vehicle. The energy density of liquid fuels does not typically pose a problem; however, the lower density of gaseous fuels under common storage technology can reduce the range.

8.3.2 Mechanical Power-Split Architectures

Mechanical power-split architectures combine power from multiple sources using a torque summing and/or speed summing mechanical coupling. The use of a planetary gear system allows for the combination of multiple mechanical linkages in these systems.

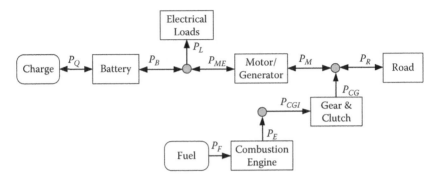

FIGURE 8.5
Parallel hybrid.

8.3.2.1 Parallel Hybrid

In a parallel hybrid, an internal combustion engine can be used to directly drive the transmission of the vehicle as seen in Figure 8.5. The power from an electric motor (P_M) through a torque summing or speed summing device can then supplement or decrement the power delivered to the road (P_R) from the combustion engine (P_{CGI}). The motor power (P_M) is chosen based on the battery SOC, and the power from the combustion engine is related to the road speed and the requirements from the chosen mechanical configuration. The power flow at the mechanical node ($P_M + P_{CGI} = P_R$) can be varied to improve the efficiency of the combustion engine.

8.3.2.2 Series-Parallel Hybrid

The series-parallel hybrid, as seen in Figure 8.6, is a combination of torque and speed coupling devices with an internal combustion engine acting as the primary energy source and multiple motors connected to a battery

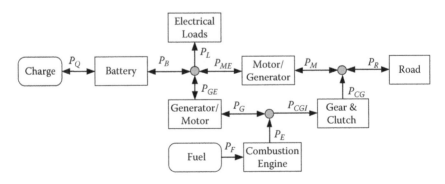

FIGURE 8.6
Series-parallel hybrid.

system acting as the secondary energy source. This architecture allows for the benefit of the series architecture and parallel architecture to be combined in a single vehicle. At low-power/low-speed operation, the vehicle can operate as a series vehicle by using a clutch to remove the parallel pathway. This allows for the road load to be met by power from the combustion engine (P_{GE}) and the battery (P_B). At high power demand/high speed, both pathways can be used such that some or all of the power from the engine (P_E) can be used at the road through the transmission (P_{CG}) and through the generator (P_{GE}) to drive the traction motor (P_{ME}). This combination is achieved through the use of a planetary gear set in which the motor/generator, combustion engine, and the linkage to the road are connected. This architecture allows for further freedom of control through the multiple pathways of power flow but requires more complex control to achieve higher efficiency.

8.3.2.3 Mechanical Power-Split Performance

The mechanical power-split architectures allow for greater flexibility in the direct use of power from combustion engines. In these designs, the motor/generator(s) are used for low power and transient requirements to augment a deliberate design reduction in peak engine power. Typically the control goal is to operate the engine in a consistent yet efficient operating point while using the electric machines to allow for increases or reduction in power at the road. This means that the acceleration times of the vehicle are a result of balancing efficient combustion engine operation with limited motor capability. Gradeability of these architectures is related to the fraction of maximum sustained power the engine can deliver through the direct and indirect power flow. Naturally the direct path will result in more efficient power; however, it is limited by the physical speed/torque relationship of the planetary gear. As a result, the gradeability of the vehicle is directly related to this gear relationship. As with the electrical power-split architectures, the range is associated with the overall efficiency of the vehicle and the capability of the fuel storage technology.

8.4 Fuel Cell Hybrid and Plug-In Hybrid Electric Vehicles

The earliest fuel cell vehicles were designed with no secondary energy storage and relied on the fuel cell for all propulsion power. These vehicles have evolved into fuel cell hybrid electric vehicles where a secondary source is added to create a series hybrid electric vehicle. The series hybrid propulsion system architectures of recent prototype vehicles and the considerations for further evolution into plug-in hybrid propulsion systems will be discussed in this section.

8.4.1 Fuel Cell Hybrid Electric Vehicles

The inherent architecture of a fuel cell vehicle is that of a series hybrid since the propulsion system of the vehicle must be of an electrical power-split type. Fuel cell systems tend to have a slower dynamic response than changes in road load require to meet expected vehicle performance. To maintain vehicle performance at least one secondary source must be used to improve the combined power response time. This power-split is achieved by the combination of at least two high-voltage power buses with the application of a power electronic converter. For fuel cell hybrids with only one secondary source, there are fundamentally only two different vehicle architectures: primary-source conversion or secondary-source conversion as shown in Figure 8.7 and Figure 8.8.

The essential factor in vehicle architecture design is within which pathway, from source to load, to place a power electronic converter. This inherently affects the overall efficiency and power capabilities of the chosen source as the characteristics are tied to the converter. The focus for fuel cell hybrid vehicles has been to minimize the losses in the pathway from the fuel cell to the road load. In these vehicles, the goal of the secondary source is to provide a small power source capable of short-term energy to improve acceleration, reduce start-up time, and allow for regenerative braking. The fuel cell and its associated balance of plant are designed to dynamically follow the power load and meet the peak power requirements of the vehicle.

The GM Fuel Cell Equinox, Honda FCX Clarity, and Toyota FCHV vehicles are three small-release prototype vehicles that represent this fuel cell hybrid electric vehicle architecture approach. Power-flow schematics of these vehicles have been included as Figure 8.9, Figure 8.10, and Figure 8.11. These vehicle designs consist of essentially the same power-flow design though converter and voltage choices are varied. The fuel cell Equinox design allows for fuel cell voltage to vary independent of the battery voltage given the buck-boost design of the converter. The FCX Clarity and FCHV use boost

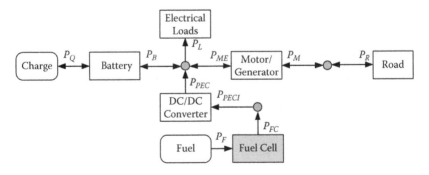

FIGURE 8.7
Primary source conversion.

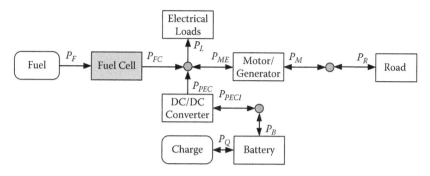

FIGURE 8.8
Secondary source conversion.

converters where the voltage range of the fuel cell is always above the battery system. The FCHV and FCX Clarity design necessitate a large number of cells in the fuel cell stack to ensure its voltage remains above the battery voltage. However, this high voltage is necessary given that all the designs utilize a stack capable of meeting peak vehicle power.

8.4.2 Fuel Cell Plug-In Hybrid Electric Vehicle

The concept of a fuel cell plug-in hybrid electric vehicle changes the main motivation for the vehicle architecture. The operational intent of the vehicle will constrain the overall efficiency of the vehicle and influence the desire for a primary- or secondary-source conversion design. The propulsion system goals are to achieve a fully operational battery electric vehicle, reduce fuel cell stack peak power, and maintain acceptable performance.

The central decision for design of a fuel cell plug-in hybrid is between primary- and secondary-source conversion. The system goal of obtaining a fully operational electric vehicle requires that in the secondary-source conversion

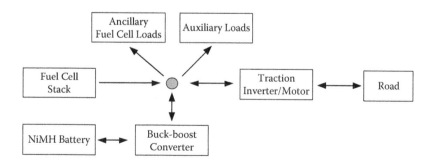

FIGURE 8.9
GM fuel cell equinox. (From Woody, G. R. et al., Automotive Power Electronics with Wide Band Gap Power Transistors, United States Patent, 2010.)

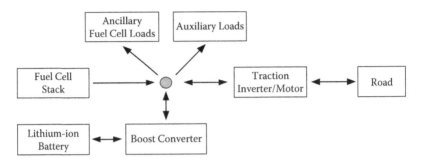

FIGURE 8.10
Honda FCX clarity. (From Kojima, Y. et al., Method of Protecting Electricity Storage Device in Hybrid dc Power Supply System, United States Patent, 2009.)

method, the main converter must be capable of the full power requirements of the traction inverter. This results in a converter power capacity of at least the sustained power rating of the traction system to maintain expected acceleration and gradeability performance of the vehicle. Secondary-source conversion selection would result in a decrease of the all-electric range due to increased pathway losses, reduced regenerative braking performance, and increased control complexity for stable operation of the fuel cell bus in hybrid operation.In the primary-source conversion case the connection of the battery directly to the traction system results in decreased losses, over the secondary-source case, for all electric or low-power charge-depleting operation. The regenerative braking performance is increased due to the removal of the converter between the source of energy, the traction system, and the energy storage system. Further, the control complexity for stable operation of the

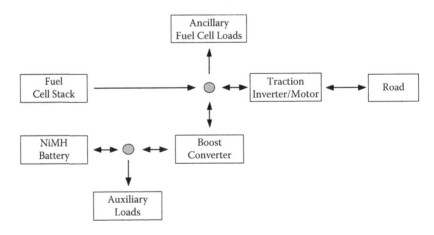

FIGURE 8.11
Toyota FCHV. (From Ishikawa, T. et al., Development of Next-Generation Fuel Cell Hybrid System—Consideration of High Voltage System, presented at the SAE 2004 World Conference Detroit, Michigan, 2004.)

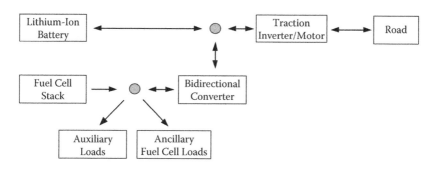

FIGURE 8.12
Auxiliary and ancillary load on primary bus.

system is reduced by placing the main load on the stiffer, lower-impedance battery bus instead of the weaker fuel cell bus. However, this method does increase the pathway losses from the primary-source resulting in a decrease in the effective efficiency of the fuel cell system. Primarily the design intent of a PHEV is to operate under battery electric or in a heavily charge-depleting regime. Subsequently, primary-source conversion architecture is assumed for the discussion of auxiliary and ancillary loads.

The placement of the remaining loads of the propulsion system can be on the secondary or primary bus. These loads are the ancillary fuel cell components and the auxiliary vehicle components such as the converter for the low-voltage system, air-conditioning compressor, and so on. Placement of these loads on the primary bus as in Figure 8.12 results in increased efficiency of the fuel cell system as the required ancillary loads are powered directly from the fuel cell. However, this design requires bidirectional power-flow through the converter for use during fuel cell start-up to power the ancillary loads. Additionally, bidirectional flow is required during battery electric vehicle operation to provide power to the auxiliary loads.

Conversely, the placement of these loads on the secondary bus of the vehicle, as in Figure 8.13, would allow for the converter design to have unidirectional

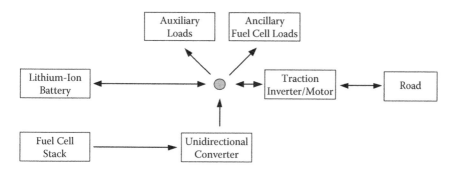

FIGURE 8.13
Auxiliary and ancillary load on secondary bus.

power-flow, reducing the converter cost and control complexity. Secondary bus placement would also limit the required operational voltage range of the load components to the smaller range of the battery system. In addition, this placement would reduce the requirement of converter operation while the fuel cell is not providing power, as during battery-electric vehicle operation.

8.5 Energy Management

Vehicle architecture selection with component sizing and energy combination method gives an inherent characteristic of attainable performance. How the vehicle control chooses to utilize these energy sources to provide power during operation will affect the achieved performance. This decision process is known as the energy management or the supervisory control of the vehicle. The management methods will depend on the chosen vehicle architecture; however, they can be broadly classified as either rule-based or optimization-based methods [9].

The common goal of all energy management methods is to capitalize on the hybrid architecture of the vehicle to generate optimal performance. However, this is often a conflicting ambition since optimization of some performance criteria results in a reduced outcome for other performance criteria. Commonly the goal is to improve fuel economy while maintaining consumer-accepted acceleration times and gradeability. More specifically, the control objective is to operate the propulsion system such that (1) the primary energy source is used in its high-efficient operating range; (2) the vehicle meets all road load requirements maintaining vehicle performance; (3) pollutants from the primary energy source are minimized; and (4) a healthy energy level in the secondary energy source is maintained [10]. In the following sections these control methods will be explored focusing on the series hybrid fuel cell vehicle architecture.

8.5.1 Heuristic Rule-Based Methods

Rule-based energy management methods use heuristics taken from power flow design, component efficiency maps, and other boundary conditions of the components to develop a deterministic set of rules for operation [9]. These methods typically use state machines to allow the control to move from different operational states based on defined rules.

8.5.2 Hysteresis (On-Off) Control

Hysteresis or ON-OFF control of a hybrid vehicle is a simple method of regulating the energy content, or state of charge (SOC), of the secondary storage system. This method operates in two basic states where the primary energy source is either in the ON or OFF state. When the SOC of the secondary source is low, the primary source is switched ON. In the ON state, the primary source provides power at a highly efficient operating point and is used to power the main traction motor (P_{ME}) and the battery (P_B). Depending on the operating point of the vehicle, the secondary source may be recharged or may provide power with the primary source to meet a high road load. Once the SOC of the secondary source has reached a high state of charge, the primary source is switched to the OFF state and all of the road load power is met by the secondary source. This process will continue keeping the SOC of the secondary source between low and high. Hysteresis control can be applied to series hybrid vehicles since the traction motor and battery source can be designed to meet the road load without the primary source. This method can suffer from inefficiency for fuel cell hybrids due to frequent changes in the ON/OFF state. The primary energy source may take a long period of operation to reach an efficient operating temperature due to the periodic nature of the control. Furthermore, frequent charging and discharging of the battery system can result in round-trip efficiency losses and decreased life of the battery system.

8.5.3 Power Follower Control

The power follower control method is a heuristic method that divides vehicle operation into identifiable modes. Fuel cell systems have parasitic loads for the balance of plant components and activation losses, which cause reduced efficiency for low power output. At higher power output the system is affected by ohmic and transportation losses. The goal of the power follower is to not only operate the fuel cell as close to the instantaneous power demand as possible but also avoid inefficient operating points. Control of the fuel cell output power is based on traction motor power requirements, vehicle speed, and state of charge (SOC) of the secondary source as depicted in Figure 8.14. The operation of the primary source is divided into four modes: IDLE, REDUCED, DIRECT, INCREASED [11–13].

8.5.3.1 Fuel Cell IDLE

Idle operation of the fuel cell occurs when the SOC of the secondary source is higher than a fixed minimum SOC set point. This requirement ensures that the secondary source has sufficient energy to operate. Once the SOC is considered high, IDLE operation occurs when either the power demand from the motor is low or the vehicle speed is low. During IDLE operation the fuel cell provides power at its highest system efficiency operating point as shown by the idealized curve in Figure 8.15. This operating point will vary for each fuel cell system but is where the balance of plant losses are overcome

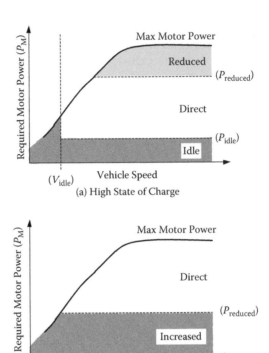

FIGURE 8.14
Power follower behavior for high and low SOC.

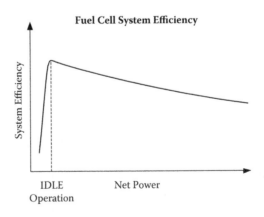

FIGURE 8.15
Idealized system efficiency vs. net power.

by the power produced. If the fuel cell system is capable of frequent start/stop operation, the system can be shut down during this mode.

8.5.3.2 *Fuel Cell DIRECT*

DIRECT operation occurs when the SOC of the secondary source is in a high or low state as shown in Figure 8.14. During DIRECT operation the fuel cell provides all of the power (P_{FC}) required for the motor (P_M) to meet the road load (P_R). The secondary source is used to meet short-term instantaneous power requirements until the ramp up or down of the fuel cell has been achieved. This allows for the motor power request to be met even though the response time of the fuel cell system is slow.

8.5.3.3 *Fuel Cell REDUCED*

REDUCED operation occurs only when the secondary source has a high SOC and the motor requires high power to meet the road load. During REDUCED operation the fuel cell provides a portion of the power required by the motor to meet the road load while the remainder of power is provided by the secondary source. This operation allows for the fuel cell to maintain operation for lower, more efficient power while using the secondary source. If the fuel cell has been sized to be smaller than the maximum power of the motor, then this operation may be seen in both high and low state (though not depicted in Figure 8.14).

8.5.3.4 *Fuel Cell INCREASED*

INCREASED operation occurs when the secondary source SOC is low and the power required to meet the road load is low. During INCREASED operation, the fuel cell provides additional power to recharge the secondary source while meeting the road load requirement. This operation allows the secondary source to provide power at a later time when the use of the fuel cell is less efficient or too slow to meet the transient power requirements of the motor.

The efficiency of the power follower method is sensitive to the boundaries set by the idle speed limit (V_{idle}), idle power limit (P_{idle}), reduced power limit ($P_{reduced}$), and increased power limit ($P_{increased}$). These parameters are used to develop the mapping of operational modes to power requirements as shown in Figure 8.14. Further, the amount by which to increase or reduce the power operation in the INCREASED and REDUCED modes is typically a fixed single value or table of values based upon vehicle speed. Tuning of the parameters is typically accomplished through intuition of how the hybrid vehicle architecture should operate.

8.6 Real-Time Optimization Methods

Ideally an optimal energy management control system should be capable of minimizing the overall fuel consumption over any given driving cycle. Therefore, minimization strategies should not merely aim to minimize fuel consumption at each instance during the drive cycle. Otherwise, the result would be a strategy that drains the secondary source. Rather, the goal is to minimize use of fuel for the entire driving cycle resulting in an over-all reduction in fuel consumption. The problem comes in formulating a minimization criterion for real-time control that does not require knowledge of the entire driving cycle *a priori*. The following control strategies make approximations to allow implementation that result in near optimal control.

8.6.1 Offline or Static Optimization

A common method of implementing optimal control for the minimization of fuel consumption is to perform a minimization on a reduced state structure offline and then apply these calculated values to a controller of similar structure. An example of this method is based on the principle of the power follower control explored in the heuristic control section. The fuel cell power $P_{FC}(t)$ is defined as an affine function of the motor power demand $P_M(t)$:

$$P_{FC}(t) = P_{FCa} + \alpha(P_M(t) - P_{Ma})$$

$$(8.1)$$

where P_{Ma} is the average power of the motor demand, while α and P_{FCa} are the fixed control parameters to be optimized offline. This strategy allows for the power of the fuel cell to be controlled based on an average fuel cell power amount (P_{FCa}) plus a fraction (α) of the current power demand ($P_M(t)$) and the average power (P_{Ma}) over the sliding average of past power demand. For $\alpha = 1$, the fuel cell will closely follow the power demand of the drive cycle and reduce the use of the secondary source; however, for $\alpha = 0$, the fuel cell will operate at the constant value of P_{FCa} and increase the use of the secondary source. The average power of the motor (P_{ma}) is calculated online and can be formed using a filtering algorithm.

The fixed control values α and P_{FCa} are supplied from an offline minimization of fuel consumption for chosen drive cycles. As a result, the optimality of the control is only as accurate as the approximation of the chosen offline drive cycles are to the online use of the control. This method will suffer from variability of the required average fuel cell power (P_{FCa}), resulting in an inability to guarantee charge sustenance of the secondary source. However, this can be overcome with a slow integrative controller for the management of the secondary source SOC [14, 15].

8.6.2 Equivalent Consumption Minimization Strategy (ECMS)

The ECMS achieves energy management control through a modified minimization of instantaneous fuel consumption. The direct minimization of instantaneous fuel consumption from the fuel cell may not yield a reduction in fuel consumption for the entire drive cycle. This is due to the inability to account for the fuel consumption represented in the charge or discharge of the secondary energy source. The ECMS reduces a global optimization problem into a local optimization problem by assigning an equivalent fuel flow rate for the use of energy from the secondary source. As a result, the total cost criterion for all time can be expressed as

$$\sum_{\{P_{FC}(t)\}} \min \quad \dot{m}_{f\_combined}(t) \qquad (8.2)$$

where the combined fuel flow rate cost function $\dot{m}_{f\_combined}(t)$ is simply the summation of the fuel flow rate of the fuel cell $\dot{m}_{f\_fc}(t)$ and the equivalent fuel flow rate for secondary source $\dot{m}_{f\_battery}(t)$.

$$\dot{m}_{f\_combined}(t) = \dot{m}_{f\_fc}(t) + \dot{m}_{f\_battery}(t) \qquad (8.3)$$

Developing the equivalent mass flow for the secondary source can be accomplished in many ways. A simple method is to calculate the average fuel consumption required of the primary source to charge the secondary source. This method takes into account the losses associated with the charging and discharging efficiencies of the secondary source. The equivalent hydrogen mass flow rate can be calculated as follows:

For positive power flow or discharging of the source:

$$\dot{m}_{f\_battery}(t) = \frac{\overline{SC}_{battery} \cdot P_B}{Eff_{dis\_battery} \cdot Eff_{pec}} \qquad (8.4)$$

For negative power flow or charging of the source:

$$\dot{m}_{f\_battery}(t) = \overline{SC}_{battery} \cdot P_B \cdot Eff_{dis\_battery} \cdot Eff_{pec} \qquad (8.5)$$

where $\overline{SC}_{battery}$ is the specific consumption that represents the amount of hydrogen needed to store electrochemical energy in the battery. The specific consumption can be represented by the following:

$$\overline{SC}_{battery} = \frac{\overline{SC}_{fc}}{Eff_{pec} \cdot Eff_{chg\_battery}} \qquad (8.6)$$

where $\overline{SC_{fc}}$ is the average specific consumption of hydrogen by the fuel cell for one kWH of converted electrical energy, $\overline{Eff_{pec}}$ is the average efficiency of the dc-dc converter, $\overline{Eff_{dis\_battery}}$ is the average discharging efficiency, and $\overline{Eff_{chg\_battery}}$ is the average charging efficiency of the battery. SOC sustenance can be ensured by weighting the fuel consumption value with a penalty function that discourages use of the secondary source when SOC is low. The following two equations describe a possible cubic-based weighting function:

$$f_{penalty} = 1 - (1 - 0.8 \cdot x_{SOC}) \cdot x_{SOC}$$

$$x_{SOC} = \frac{SOC - \dfrac{(SOC_L + SOC_H)}{2}}{(SOC_H - SOC_L)} \tag{8.7}$$

where SOC_H and SOC_L are the high and low boundary conditions for the charge-sustaining operation window.

The selection of an operating point can be achieved in a time-discrete manner with an online algorithm. This algorithm starts with the power required by the motor (P_M) to meet the road load. The motor power is then broken into several candidate power requests for the fuel cell (P_{FC}) and the battery (P_B). The fuel flow rates for each candidate is determined and the candidate with the minimum combined fuel flow rate ($\dot{m}_{f\_combined}(t)$) is selected. The mass flow rate of the fuel cell can be approximated using predefined tables that take into account the operating temperature and balance of plant losses.

ECMS allows for the optimization problem to be solved in real-time; however, the evaluation of the equivalent fuel consumption introduces an approximation. The method described here relies on the average value calculations, which can distort the true cost of the energy stored in the secondary source. Adaptations of this method have been shown to improve these approximations, resulting in improved performance, though the concept of the method remains the same as the one presented here [9, 15–17].

References

[1] A. Emadi, et al. 2005. Topological overview of hybrid electric and fuel cell vehicular power system architectures and configurations. *IEEE Trans. Veh. Technol.*, vol. 54, pp. 763–770.

[2] M. Ehshani, et al. 2010. *Modern Electric, Hybrid Electric, and Fuel Cell Vehicles: Fundamentals, Theory, and Design*, 2nd ed. Boca Raton, FL: CRC Press.

[3] M. Douba. 2009. Argonne Facilitation of PHEV Standard Testing Procedure (SAE J1711).

[4] G. Beauregard. 2004. HICE Vehicle Acceleration, Gradeability, and Deceleration Test Procedure, in *Hydrogen Internal Combustion Engine Vehicle Specificationas and Test Procedures*, ed: Electric Transportation Applications.

[5] C. C. Chan. 2007. The State of the Art of Electric, Hybrid, and Fuel Cell Vehicles. *Proc. IEEE*, vol. 95, pp. 704–718.

[6] G. R. Woody, et al. 2010. Automotive power electronics with wide band gap power transistors. United States Patent.

[7] Y. Kojima, et al. 2009. Method of protecting electricity storage device in hybrid dc power supply system. United States Patent.

[8] T. Ishikawa, et al. 2004. Development of next-generation fuel cell hybrid system—Consideration of high voltage system. Presented at the SAE 2004 World Conference Detroit, Michigan.

[9] F. R. Salmasi. 2007. Control Strategies for Hybrid Electric Vehicles: Evolution, Classification, Comparison, and Future Trends. *IEEE Trans. Veh. Technol.*, vol. 56, pp. 2393–2404.

[10] L. Guzzella and A. Sciarretta. 2007. *Vehicle Propulsion Systems: Introduction to Modeling and Optimization*. Berlin Springer-Verlag.

[11] J. Bauman and M. Kazerani. 2008. A Comparative Study of Fuel-Cell-Battery, Fuel-Cell-Ultracapacitor, and Fuel-Cell-Battery-Ultracapacitor Vehicles. *IEEE Trans. Veh. Technol.*, vol. 57, pp. 760–769.

[12] W. Di and S. S. Williamson. 2007. Performance Characterization and Comparison of Power Control Strategies for Fuel Cell Based Hybrid Electric Vehicles. Presented at the Vehicle Power and Propulsion Conference.

[13] V. H. Johnson, et al. 2000. HEV control strategy for real-time optimization of fuel economy and emissions. Presented at the Future Car Congress, Arlington, VA.

[14] S. Barsali, et al. 2004. A control strategy to minimize fuel consumption of series hybrid electric vehicles. *IEEE Trans. Energy Convers.*, vol. 19, pp. 187–195.

[15] A. Sciarretta and L. Guzzella. 2007. Control of hybrid electric vehicles. *IEEE Control Syst. Mag.*, vol. 27, pp. 60–70.

[16] Y. G. G. Paganelli, and G. Rizzoni. 2002. Optimizing control strategy for hybrid fuel cell vehicle. *SAE Transactions*, vol. 111.

[17] L. Serrao, et al. 2009. ECMS as a realization of Pontryagin's minimum principle for HEV control, in *American Control Conference, 2009. ACC '09.*, pp. 3964–3969.

9

Hydrogen as Energy Storage to Increase Wind Energy Penetration into Power Grid

Raquel Garde, Gabriel García, and Mónica Aguado

CONTENTS

9.1 Introduction

In the last 10 years, wind power has increased its installed capacity dramatically from 5 GW in 1995 to the current 120 GW of installed capacity, 65 GW of which being installed in the United States, Germany, and Spain. This means that there are countries with a very high penetration of wind energy and others with very little or no wind energy. Even within the countries with a big share of wind, the situation can vary depending on the energy mix and the interconnection capacities. For example, Denmark, with a higher level of wind penetration, 20%, is also highly interconnected to its neighbors Norway, Sweden, and Germany, being able to export or import 100% of its peak production/load. Contrarily, Spain has only around 2.5% of interconnection capacity with the UCTE through France, needing to solve any problem due to wind on its own (in some cases, Spain and Portugal are considered together as a single system, both sufficiently interconnected and both with a high wind penetration).

Depending on its own system characteristics, energy mix, and interconnection capacity, each country chooses its own way of integrating wind energy and each case should be studied independently.

To achieve the renewable energy integration goals, it is necessary to consider the use of energy storage systems that allow a proper management of the intermittent and variable power generation based on renewables, mainly wind power.

In this work, an analysis of the wind and hydrogen full integration concept is developed given the noticeable synergy between hydrogen and wind energy regarding their integration for a further approach of renewable energy sources with inherent noncontinuous and random nature.

9.2 Wind Power Deployment

According to the study developed by the EWEA, Pure Power [1], the current electricity supply structure in Europe maintains the characteristics of the time in which it was developed. Each country has its own energy system, the technologies applied are aging, and the markets supporting it are underdeveloped. Europe needs a good internal market for electricity to face the global challenges of climate change taking into account the depleting of indigenous energy resources, the increase in fuel costs, and the threat of supply disruptions.

The estimations are that by 2020, 332 GW of new electricity capacity—42% of current EU capacity—needs to be built to replace old power plants and meet the expected increase in demand. The 2009 EU Renewable Energy Directive aims to increase the share of renewable energy in the EU from 8.6% in 2005 to 20% in 2020 and as the cheapest technology and most developed, onshore wind will be the largest contributor to meeting the 34% share of renewable electricity needed by 2020 in the EU.

EWEA has analyzed the wind power market in the 27 Member States to provide estimations about the deployment of wind energy in Europe by 2020 and 2030. EWEA has considered two scenarios, low and high. The first one is more conservative and assumes a total installed capacity in EU by 2020 of 230 GW, producing 580 TWh of electricity. The second one assumes a total installed wind power capacity of 265 GW by 2020, producing 681 TWh of electricity.

By 2030, in the conservative scenario, EWEA expects 400 GW of wind energy capacity in EU-27, with 250 GW of onshore and 150 GW of offshore.

That means that by 2030, wind power in Europe will produce 1155 TWh, meeting between 26% and 34% of the European electricity demand, depending on the demand considered as reference.

By the end of 2008, there was 64.9 GW of wind power capacity installed in the EU-27, of which 63.9 GW was in the EU-15 and the 62.6% concentrated in two countries, Germany and Spain, with 23.9 GW and 16.74 GW respectively.

According to the AWEA (American Wind Energy Association), during the third quarter of 2011 over 1200 megawatts (MW) of wind power capacity were installed, bringing installations through the first three quarters of the year to 3360 MW. The U.S. wind industry now totals 43.5 GW of cumulative wind capacity (September 2011). The U.S. wind industry has added over 35% of all new generating capacity over the past four years, second only to natural gas, and more than nuclear and coal combined. Today, U.S. wind power capacity represents more than 20% of the world's installed wind power.

In May 2008, the U.S. Department of Energy released a report, *20% Wind Energy by 2030: Increasing Wind Energy's Contribution to U.S. Electricity Supply* [2]. It concluded that the United States possesses sufficient and affordable wind resources that would enable the nation to obtain 20% of its electricity from wind.

In the year 2010, the wind capacity reached worldwide 196.6 GW and China accounted for more than half of the world wind market for new wind turbines adding 18.9 GW, which equals a market share of 50.3%. At the end of 2010, China had a wind power installed of more than 44 GW followed by the United States with 40 GW.

The countries with the highest wind shares are Denmark (21%), Portugal (18%), Spain (16%), and Germany (9%). In China, wind contributed 1.2% to the overall electricity supply, while in the United States the wind share has reached about 2% in 2010.

Based on the current growth rates, WWEA (World Wind Energy Association) expects a global capacity of 600 GW in 2015. By the end of year 2020, at least 1500 GW can be expected to be installed globally.

The IEA [3] expects 4528 GW of electricity-generating capacity to be installed worldwide in the period 2007–2030, requiring investments of $5034 billion in generation, $2106 billion in transmission grids, and $4657 billion in distribution grids.

9.3 Impact of High Wind Energy Penetration on Power Systems

Wind energy has some special characteristics that may have a significant effect on the power system, the most important being its intermittency and variability. The operation and capabilities of wind energy should be adapted to the system and this one should recognize the specific characteristics of the wind energy and adapt its operation and management rules to better integrate this new major energy source.

The impact of wind energy on the Power System affects to the operational security, reliability, and efficiency and could be analyzed from a technical and economical point of view. The impacts can be related to three focus areas: balancing, adequacy of power, and grid as the IEA WIND task 25 group has shown in some results obtained from this international collaboration [4].

9.3.1 Balancing Requirements

The current system operation is designed to follow load fluctuations at every moment and balancing mechanisms, being different kind of reserves as frequency containment reserves (FCR), frequency restoration reserve (FRR), and replacement reserve (RR), are based on forecasting values of wind power and load. This means that power station outages, stochastic load variability, and fluctuations of wind power injections are the main factors to take into account for controlling and balancing power.

The development of models for dynamical forecast uncertainty estimation as for wind power forecast are prioritized areas due to their applications concerning decision-making problems related to allocation of balancing power and reserve requirements, schedules of power generators, and bidding strategies in electricity markets, among other issues. Therefore, the accuracy on forecasting wind energy influences the conventional capacity operation as well as the unit commitment.

9.3.2 Adequacy of Power

To estimate the required generation capacity, mainly during peak load situations, system load demand and maintenance of production units are taken into account with criteria as the loss of load expectation, the loss of load probability, and the loss of energy expectation.

In power systems with increasing wind energy penetration and bigger power ratings, wind turbines and wind farms should ensure the frequency stability of the system or fulfill the power and frequency control requirements. Thus, the future wind generation should provide ancillary services that includes the provision of primary, secondary, and inertial energy (spinning reserve).

9.3.3 Grid

Wind power affects power flow in the grid and has a relevant influence in the voltage stability of the grid. The impact on the transmission network depends on the situation of wind farms relative to load and the correlation between wind power generation and energy consumption.

Wind turbine technology has been adapted to voltage stability requirements from the grid and manufacturers provide their systems with power quality filters and FACTS or STATCOMS to control reactive power and to fulfill the operator requirements of voltage quality.

Also the wind farms are obliged to provide capabilities against fault ridethrough to maintain the grid stability in case of faults. Some codes require grid support with voltage dips of 0% for different times, and many of the current wind turbines are already equipped with this specification.

In summary, in recent years, the increase of wind energy connected to the power grid has been a challenge for power system operators and electricity markets. Nevertheless, a major integration of wind energy is possible if a mutual adaptation between system operation rules and new technical and economic capabilities is established.

Wind promoters and manufacturers have worked very hard to overcome most of the handicaps stated by the system operator (fault ride-through capabilities, voltage and frequency control capabilities, or ancillary services).

Also the system operator can control a complete collection of RES plants through control centers, and improvements in forecasting wind generation can allow a better integration of wind energy in the electricity market.

However, in spite of all the improvements developed in the wind power systems and in the grid to adjust the increased wind power share, some maladjustments can occur, leading to a difficult management of the system and wind curtailments with the corresponding lost of energy.

9.4 Energy Storage Systems

9.4.1 Introduction

Energy storage is a well-established concept yet still relatively unexplored. Storage systems such as pumped-hydroelectric energy storage (PHES) have been in use since 1929 [5] primarily to level the daily load on the network between night and day. As the electricity sector is changing, energy storage is starting to become a realistic option for the following [6]:

1. Restructuring the electricity market
2. Integrating renewable resources
3. Improving power quality
4. Aiding shift toward distributed energy
5. Helping network to operate under more stringent environmental requirements

The main applications can be classified as bulk energy storage, for the purpose of load-leveling or load management, distributed generation (DG) for peak shaving, and power quality (PQ) or end-use reliability.

Energy storage can optimize the existing generation and transmission infrastructures while also preventing expensive upgrades. Energy storage devices can manage the power fluctuations from renewable resources, thus aiding the use of several renewable technologies and their large-scale penetration into the network. In relation with conventional power production, energy storage devices can improve overall power quality and reliability, which is becoming more important for modern commercial applications. Finally, energy storage devices can reduce emissions by aiding the transition to newer, cleaner technologies such as renewable resources and the hydrogen economy and distributed generation.

The different applications are distinguished by the power level and discharge time required. These specifications determine the stored energy requirements. Short-term, high-power energy storage provides reliability and power quality for the digital economy where outages of a few cycles can lead to costly downtime. Long-term energy storage can mitigate the effects of an overburdened electricity grid by peak shaving and load shifting, while reducing energy costs for consumers.

The main requirements for different applications are collected in Table 9.1.

Nevertheless, a number of obstacles have hampered the commercialization of energy storage devices including the following:

1. Lack of experience and development state of technologies
2. Inconclusive and difficult benefits quantification in terms of savings and also power quality
3. High capital costs due to, among other causes, the low market
4. Involvement for energy storage by developers or transmission system operator (TSO)

However, as renewable resources and power quality become increasingly important, costs and concerns regarding energy storage technologies are expected to decline.

9.4.2 Energy Storage Technologies

There are a number of energy storage technologies that can help support the electric power industry. Taken together, electrochemical batteries and capacitors, electromechanical systems as flywheels, superconducting magnetic devices, pumped hydro, and others can support a wide range of electric power

TABLE 9.1

Requirements for Energy Storage Application

Application	Typical Power Rating	Discharge Time	Energy Delivery	Frequency of Use	Benefit
Transmission Support	100 MW	10 s	100 MW per pulse	1/month	$50–150/kW/yr
Area Regulation	20 MW	Continuous	2 MWh	Continuous	$700–1500/kW/yr
Spinning Reserve	20 MW	15 min.	2 MWh	1/month	$700–1500/kW/yr
Transmission Facility Deferral	10 MW/15 MVA	>5 hrs.	50 MWh	100–200 days/yr	$50–150/kW/yr
Renewable Energy Management	5 MW	1–10 hrs.	1–10 MWh	10–20 days/month	$1000–1500/kW/yr
Commodity Storage	2 MW	3–4 hrs.	3–4 MWh	6/yr to daily	$120–250/kW/yr
Commercial Load Following	200 kW	3 hrs.	75–100 kWh	Daily	$10–20/kW/month

Source: Sandia National Lab.

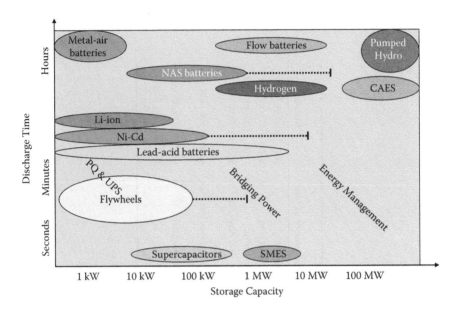

FIGURE 9.1
Ratings of energy storage systems. (Adapted from the Electricity Storage Association.)

applications [7]. In Figure 9.1 the ratings of energy storage systems can be observed [8,9].

9.4.2.1 Hydrogen

Hydrogen is probably the most immature but also one of the most promising energy storage techniques available. Hydrogen, however, is not a system but an energy carrier, so it displays numerous applications in addition to the storage of energy (e.g., hydrogen is a gas, long used as chemical in several industries as petrochemical, food, pharmacological, etc.). This gas must be generated, because is not a natural resource or primary energy; it is stored in different ways and can be used as chemical, fuel, or energy storage system.

There are many options to produce hydrogen from fossil fuels and from renewable energies, mainly by water electrolysis from wind power. It is possible to store hydrogen as compressed gas, cryogenic liquid, or even in solid state with hydrides, and hydrogen can be reconverted into electricity or thermal energy with gas turbines and internal combustion engines (ICE) adapted for hydrogen and with the especially developed technology, the fuel cells.

Here, we have focused our analysis in the following hydrogen chain: water electrolysis from wind power (in different scales), storage as compressed gas, and reconversion into electricity with ICEs or fuel cells. The application of hydrogen in the transport sector is also being analyzed as an option to enhance the added value of hydrogen, since in the transport sector hydrogen

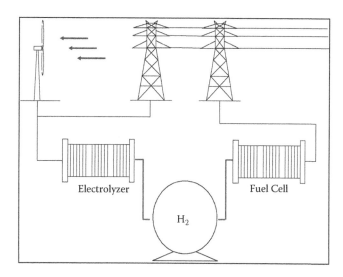

FIGURE 9.2
Scheme of an energy system based on renewable energies and hydrogen. (From CENER.)

competes with fossil fuels with high prices, which are very dependent on the rising oil prices, while energy storage competes with the kWh prices, which are much lower than those of fossil fuels.

Hydrogen, however, can be a good system to increase and improve the penetration of wind power into the electrical grids avoiding technical and management concerns. Therefore, it looks reasonable to analyze hydrogen as an energy storage system compared to other ESSs which compete in the same conditions, that is, in the electrical chain instead of transport applications.

We are not going to describe the different hydrogen technologies since that description is broadly available in the literature [10,11,12,13,14]. Nevertheless we are going to briefly describe the main characteristics of the whole system and the application when integrated with wind farms.

An electrolyzer breaks water into hydrogen and oxygen. The oxygen usually is dissipated into the atmosphere and the hydrogen is stored so it can be used for future generation. Due to the high cost of electrical production, the most attractive option is integrating electrolyzer units with renewable resources such as wind or solar. To achieve this, an electrolyzer must be capable of operating with high efficiency, under good dynamic response, over a wide input range, and under frequently changing conditions.

Recently, a number of advancements have been made including higher efficiencies of 85% (DC, stack), wider input power capabilities, and more variable inputs. Electrolyzers are modular devices so the capacity of a device is proportional to the number of cells that make up a stack, enabling to connect several stacks in series or parallel to achieve the correct voltage. There are commercial systems available which can produce until 485 Nm³/h, corresponding to an input power of 2.5 MW. The lifetime of an electrolyzer is difficult to predict due to its limited experience operating in these conditions. Nevertheless, the standard electrolyzers have been revealed to possess lifetimes longer than 20 years.

Hydrogen can be compressed into containers or underground reservoirs. Compression is a relatively simple technology, but the energy density and efficiency (65% to 70%) are low. Also, problems have occurred with the mechanical compression. However, this is at present the most common form of storing hydrogen and for the transport industry the pressure rises to approximately 700 bar. However, the energy required for the compression is a major drawback.

Regarding the reconversion technologies, it is expected that the ICE will act as a transition technology while fuel cells are improving, because the modifications required to convert an ICE to operate on hydrogen are not very significant. However, the FC, due to its virtually emission-free, efficient, and reliable characteristics, is expected to be the generator of choice for future hydrogen-powered energy applications.

A fuel cell converts the stored chemical energy as hydrogen, directly into electrical energy. There are several fuel cells that can be classified by the electrolyte used or the operating temperature: PEM fuel cell (proton exchange membrane), alkaline fuel cell (AFC), phosphoric fuel cell (PFC), molten carbonate fuel cell (MCFC), and solid oxide fuel cell (SOFC). PFC, MCFC, and SOFC operate at high temperature and could profit the heat for cogeneration increasing the overall efficiency to 85% while the PEMFC and AFC are more efficient from the electrical point of view (40–45%); however, the CHP applications are still under development. The two main drawbacks of these systems are the high costs and the short lifetimes.

9.4.3 Comparative between Energy Storage Systems

There are many parameters describing the characteristics and adequacy of the different energy storage systems to the different energy applications. Among them, we can mention size, time of storage, charge and discharge times, costs, and so on. Therefore, it is difficult to compare the ESSs and, to determine which is more convenient for an application, it is necessary to analyze several parameters related to that specific application.

Nevertheless it is possible to categorize the ESSs depending on the range of discharge time on very short-term (<1 min), short-term (< 2 hours), long-term (2–8 hours), and very long-term (days or weeks).

In Table 9.2 we collect the main energy applications and the adequate technologies to achieve the objectives according to the study developed by Sandia National Laboratories [15]. Figure 9.3 shows the characteristics of the different energy storage systems according to their energy and power capabilities.

As can be seen, flywheels, supercapacitors, and SMES are better adapted to very short-term and short-term applications, while compressed air energy

TABLE 9.2

Applications and Appropriate Technology

Application	Power	Storage Time	Energy kWh	Response Time	Technologies
Very short-term					
End-use ride-through, Power Quality, Motor Starting	≤ 1MW	secs	~0.2	< 1/4 cycle	Flywheel Supercapacitor Micro-SMES Lead-acid battery Flow battery H2 fuel cell
Transit	< 1MW	secs	~0.2	< 1 cycle	Flywheel Supercapacitor Micro-SMES Lead-acid battery Flow battery H2 fuel cell
T&D stabilisation	up to 100s MW	secs	20–50	< 1/4 cycle	SMES H2 fuel cell Lead-acid battery Flow battery
Short-term			**kWh**		
Distributed generation (peaking)	0.5–5 MW	~1 hour	5,000–50,000	< 1 min	Flywheel Advanced battery SMES Lead-acid battery Flow battery Fuel cell or engine

(continued)

TABLE 9.2 (CONTINUED)

Applications and Appropriate Technology

Application Very short-term	Power	Storage Time	Energy kWh	Response Time	Technologies
End-use peak shaving (to avoid demand charges)	<1 MW	~1 hour	1,000	<1 min	Flywheel Advanced battery Lead-acid battery Flow battery SMES Fuel cell or engine
Spinning reserve – rapid response within 3 sec to avoid automatic shift	1–100 MW	<30 min	5,000–500,000	<3 sec	Flywheel Lead-acid battery Advanced battery Flow battery SMES Fuel cell or engine
Conventional – respond within 10 min	1–100 MW	<30 min	5,000–500,000	<10 min	Flywheel Lead-acid battery Advanced battery Flow battery SMES Fuel cell or engine CAES Pumped hydro
Telecommunications backup	1–2 kW	~2 hours	2–4	<1 cycle	Flywheel Supercapacitor Lead-acid battery Advanced battery Flow battery H2 fuel cell

Application					
Renewable matching (intermittent)	Up to 10 MW	Min 1 hour	10–10,000	<1 cycle	Flywheel, Lead-acid battery, Advanced battery, Flow battery, H2 fuel cell, SMES
Uninterruptible Power Supply	Up to 2 MW	~2 hours	100–4,000	secs	Flywheel, Lead-acid battery, Advanced battery, Flow battery, SMES, H2 fuel cell, H2 engine
Long-term			**MWh**		
Generation, load levelling	100's MW	6–10 hours	100–1,000	mins	SMES, Lead-acid battery, Advanced battery, Flow battery, Pumped hydro, CAES, H2 fuel cell, H2 engine
Ramping, load following	100's MW	Several hours	100–1,000	< cycle	SMES, Lead-acid battery, Advanced battery, Flow battery, H2 fuel cell

(continued)

TABLE 9.2 (CONTINUED)

Applications and Appropriate Technology

Application Very short-term	Power	Storage Time	Energy kWh MWh	Response Time	Technologies
Very long-term					
Emergency backup	1 MW	24 hours	24	secs–mins	Lead-acid battery Flow battery H2 engine H2 fuel cell Advanced battery
Seasonal storage	50–300 MW	weeks	10,000–100,000	mins	CAES Flow battery Advanced battery CAES Pumped hydro H2 fuel cell with Underground storage
Renewables backup	100 kW–1 MW	7 days	20–200	sec–mins	

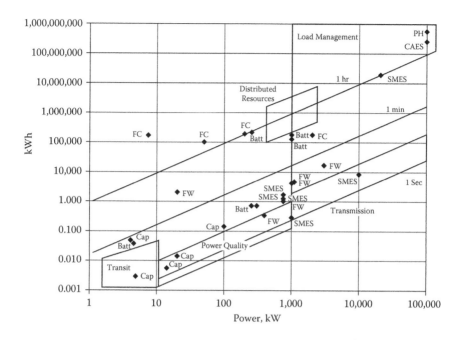

FIGURE 9.3

Power and energy characteristics of energy storage systems. (FW = Flywheel, SMES, Batt = Lead battery, Cap = supercapacitor, CAES = Compressed Air, PH = Pumped Hydro, FC = Fuel Cell). (From Schoenung, S. M., *Characteristics and Technologies for Long- vs. Short-Term Energy Storage*, Sandia Report, SAND2001-0765, March 2001; Schoenung, S. M. and W. V. Hassenzahl, *Long- vs. Short-Term Energy Storage Technologies Analysis—A Life-Cycle Cost Study*, Sandia Report, SAND2003-2783, August 2003.)

storage (CAES) and pumped hydro are better for very long-term applications and in some cases for short-term high-energy applications. The main applications are related to load management and emergency reserves.

Batteries in general are well adapted to any application as well as hydrogen technologies, taking into account that electrolyzers and fuel cells are also electrochemical devices. The applications go from power quality to load leveling or distributed generation. According to that, hydrogen could cover the whole range of energy applications and even some power applications competing with flywheels or supercapacitors. However, these applications require a further development of the FC technology. Nevertheless, the hydrogen use is not broadly spread due to two main factors, the high costs and the low cycle efficiency.

According to Table 9.2 the most appropriate applications for hydrogen are probably the short-term and long-term applications linked to renewables depending on the energy management strategy selected for the system.

If we take into account the commercial maturity, costs, and efficiency of the different energy storage systems, we observe that hydrogen is not well placed compared to other ESSs. In Table 9.3 we collect the main data related

TABLE 9.3

Comparison of Different Parameters for Energy Storage Systems [15, 16]

Technology	Energy-Related Cost ($/kWh)	Power-Related Cost ($/kW)	Balance of Plant ($/kWh)	Electrolyzer ($/kW) & Compressor ($/scfm)	Cost Certainty	Commercial Maturity	Discharge Efficiency	Lifetime (years)
Lead-acid batteries	200–400	300–600	50		1	1	0.85	6
Advanced batteries	100–2,500	150–4,000	40		3–4	2–3	0.70	10
Flow batteries	150–1,000	600–2,500	30–50		2	2–3	0.70	10
Micro-SMES	72,000	300	10,000			2	0.95	20
SMES	1,000–10,000	200–300	100–1,500		2	2	0.95	20
Flywheels (HS)	25,000	350	1,000		3	3	0.93	20
Flywheels (LS)	300	280	80		1	2	0.90	20
Supercapacitors	300–2,000	100–300	10,000		3	3	0.95	20
CAES	2–50	400–800	50		2	2–3	0.79	30
Pumped Hydro	5–100	600–2,000	2		1	1	0.87	30
H2 FC/Gas Storage (Low)	15	3,000–6,000	50	300+112.5	2–3	2–3	0.59	10
H2 FC/Gas Storage (high)	15	4,500–7,500	50	600+112.5	2–3	2–3	0.59	10
H2FC/Underground Storage	1	3,000–6,000	50	300+112.5	3	3	0.59	10
H2 ICE/Gas Storage	15	2,000–2,500	40	300+112.5	2	2–3	0.44	10

Legend for the Table

Number	Commercial Maturity	Cost Certainty
1	Mature products, many sold	Price list available
2	Commercial products, multiple units in the field	Price quotes available
3	Prototype units in the field	Costs determined each project
4	Designs available	Costs estimated

Source: Compiled from Schoenung, S. M., *Characteristics and Technologies for Long- vs. Short-Term Energy Storage*, Sandia Report, SAND2001-0765, March 2001; Schoenung, S. M. and W. V. Hassenzahl, *Long- vs. Short-Term Energy Storage Technologies Analysis, A Life-Cycle Cost Study*, Sandia Report, SAND2003-2783, August 2003; and Chen, H., T.N. Cong, W. Yang, C. Tan, Y. Li, and Y. Ding, Progress in Electrical Energy Storage System: A Critical Review, *Progress in Natural Science*, 19, 2009, 291–312.

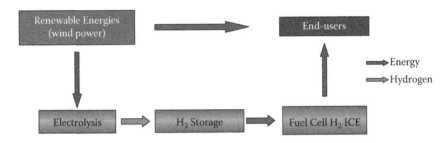

FIGURE 9.4
Scheme of energy isolated system based on renewable energies and hydrogen technologies as energy storage. (From CENER.)

to these parameters for the different ESSs analyzed in this study. Most of the costs have been calculated based on a mass production for the system.

Regarding commercial maturity, hydrogen technologies are at the same level as other competitors as batteries, except for lead-acid batteries, which are the only technology completely developed and commercialized. Nevertheless, the integration of hydrogen with renewable energies attracted the attention of many potential investors some years ago and there are many demonstration projects that provide the technical feasibility of these systems.

Regarding the costs, hydrogen technologies are better suited for energy than power applications as we can see in the table. However, in both applications, efficiency is a key factor and hydrogen fuel cells and ICEs are much less efficient than batteries in the short-term, and less than pumped hydro in the long-term applications, for instance.

On the other hand, if we also take into account the lifetime, hydrogen technologies are well placed compared with other competitors, mainly lead-acid

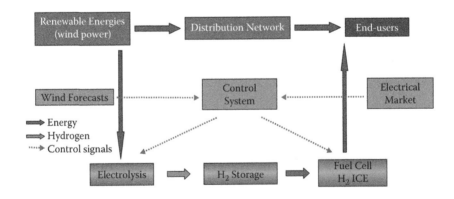

FIGURE 9.5
Scheme of distribution energy system based on renewable energies and hydrogen. (From CENER.)

batteries. The short lifetime of the lead batteries could compensate the higher costs of other electrochemical systems and, in the case of hydrogen technologies compared to other batteries, the storage time capacity without losses is another advantage to be included in the analysis.

If we now consider the very long-term applications, the main competitors of hydrogen are CAES and pumped hydro systems. Both of them are less expensive than hydrogen technologies, the efficiencies are higher than the hydrogen technologies efficiency, and pumped hydro is completely developed; in fact, it is the only current energy storage system widely used.

However, the main drawback of these systems is the environmental impact, since these facilities require natural emplacements that are not always available; generally, environmental impact studies are needed, as well as permits that entail long time procedures to be obtained. In those cases, hydrogen technologies are easier to install and costs and lower efficiency could be compensated.

9.4.3.1 Energy Management Strategies in Energy Storage Applications

As mentioned before, hydrogen is an energy vector that is very versatile and useful in different sectors. In energy applications, there are several options since hydrogen can be used in the following:

- Portable applications, small electronic devices, backup systems, UPS, telecom applications, etc.
- Stationary applications, residential and households systems, distributed generation and smart-grids, energy storage, etc.
- Transport sector as fuel for automotive applications

Focused on energy storage applications linked to wind power, the strategy adopted to manage the energy in the wind farm or in the wind power system is a main factor to be considered in the analysis.

We have described the main applications of energy storage systems and, depending on the scale, hydrogen technologies can be applied associated to wind power for load leveling (peak shaving) at small (wind farm) and large scale (global grid), for reduction of forecasts deviations, peaking plants, and repowering.

Load leveling at wind farms or at the substation level (for a local group of wind farms) basically consists of managing the wind power according to the demand profile at this point. In other words, when the demand is low and there is an excess of wind power production, hydrogen is produced and stored. When the demand is high, it is covered by the wind power and the hydrogen if necessary [16]. This kind of system is usually an isolated systems at small scale, and when the scale is higher, generally wind power is supported by other power generation systems and the hydrogen system needed is of few hundreds of kWs maximum.

It can also be applied in distributed generation systems based mainly on renewable energies and connected to distribution grids, which should basically satisfy the load curve in a medium scale of several MWs maximum.

The main competitors in these systems are the batteries that can follow the load profile as hydrogen technologies, because they are also electrochemical devices but their efficiencies are higher.

The load leveling at large scale could be described as a demand management or virtual energy storage. Electrolyzers can be considered as simple loads placed in different sites and with different sizes, in gas stations, in substations, in houses, or in centralized plants [17,18]. Depending on the wind power generation available, the rest of the foreseen demand, and the rest of the power generation system, the electrolyzers will either work or not absorbing the excess of energy to produce hydrogen but, in this case, the gas could not be used to produce electricity but as fuel for transport applications. The final application of this hydrogen will depend on the price of the competitors, the kWh or the gasoline liter.

In this kind of application, electric vehicles constitute a very important competitor for hydrogen, as EV can be charged during valley hours (night) and used during the day for transport and, even in the V2G concept, as support during peak hours giving back electricity to the grid.

The *forecast deviation* reduction is another application for hydrogen. In that case, the operator of the wind farm acquires some commitments on energy sales based on the wind forecasts. If the commitments are not satisfied because the wind power differs from the forecast, the operator must pay some penalties. To avoid that, hydrogen is produced when the generation excesses the compromises and is transformed into electricity to cover the deficit when wind generation is below the compromises. In this application the size of the hydrogen plant can be quite reduced, but it depends on the average deviation of the farm, taking into account that in some emplacements the forecasts deviations could rise up to 20%.

According to the Spanish Tariffs Regulation, the wind power promoters have two options to sell the energy: (1) with a regulated tariff where the deviation penalties are subtracted from the final price of the kWh or (2) the usual way, in the electricity market through an operator who manages several wind farms, reducing the deviations among them. The penalties are defined depending on the final tariff of the electricity achieved according to the power generation planning.

In Spain, the penalties are not high enough to compensate the energy losses due to the low efficiencies and the high costs of the hydrogen technologies. Moreover, there is no regulation about the tariff of the electricity produced by ICEs, which should be higher with bonus to make the systems cost-competitive.

The main competitors in this application are also the batteries, and more concretely the flow batteries due to their efficiency and high capability of cycling and, as for hydrogen technologies, the power and energy capacities are independent.

The *peaking plants* are a very interesting application for hydrogen, in which the gas is produced during the night (valley hours when electricity is cheaper) and used as electricity during one or two peak hours per day when the electricity price is higher. The higher the gap between the prices per kWh of peak and valley hours is, the more cost-competitive this strategy will be.

Depending on the country and its generation mix, the gap between prices will be higher or lower. Usually when the hydropower generation is high, the prices stabilize and there is a reduction of the gap between the peak and the valley tariffs due to the flexibility of managing of the hydropower plants.

Therefore, the peaking strategy is more cost-competitive in fossil fuel–based markets with more temporary fluctuations in the electrical tariffs. In Spain, the power generation is based mainly on fossil fuels and this strategy is well suited for hydrogen. However, the growing wind power penetration in the market is also stabilizing the tariffs (as hydropower) and the gap between prices is still not enough to compensate the low efficiency and high costs of the hydrogen technologies.

As we mentioned above, hydrogen is well suited to this application since the capabilities of the electrolyzer and the reconversion systems are independent. That permits designing a hydrogen plant with medium-size electrolyzers, large-size fuel cells or ICEs, and quite small hydrogen storage, given that the system has a daily cycling operation.

The only main competitor in this application could be flow batteries since the power and energy capabilities are also independent. However, in flow batteries the stack used in the charge and discharge is the same and in spite of the possibility to discharge twice or three times the rated power, the time of discharge in these conditions is limited.

Finally, the repowering application [19] makes sense in countries as Spain, Germany, or Denmark where the technology of some of their installed wind turbines is outdated or the best emplacements are underexploited.

In Spain, for example, the rated power of the wind farms cannot overcome the 5% of the power in the PCC (point of common coupling). Therefore, the rated power of the wind farm will be as high as this value. Most of the best emplacements are now exploited with wind turbines of rated power lower than 1 MW while the new technologies are all over this value. If we want to increase the wind farm capacity factor in the PCC, we can increase the rated power of the wind farm but several hours per year the generation will exceed the PCC limit and the wind farm will be disconnected, losing energy and benefits. Hydrogen can be produced by electrolysis during these hours, stored, and finally used to produce electricity during low wind generation, increasing the capacity factor, or used as fuel for transport if possible.

In this application the competitors can be batteries and also CAES and pumped hydro depending on the size of the wind farm (offshore wind farms are much larger than onshore wind farms with rated power up to 800 MW).

9.4.3.2 Hydrogen Production Centralized vs. Decentralized

Taking into account the stationary applications, at present there are two main systems that can use H_2 as electricity storage system with wind power:

- Off-grid or isolated systems, whose main objective is to satisfy the load of the community in an isle configuration [20,21,22,23,24,25,26].
- On-grid systems whose main objective is to manage and optimize the wind power generation increasing the capacity factor. This approach allows taking advantage of the generation peaks to produce hydrogen and using it during deficit periods [27]. Moreover, in this way it is possible to operate with low tariffs to produce hydrogen [28].

Both approaches must be analyzed from different perspectives since the objectives as well as the operation strategies are different. Moreover, the development of this kind of facilities depends on the countries, regulations and the grid capacity regarding the renewable energies and their management [29].

Hydrogen can be cost-competitive in isolated and remote emplacements where fossil fuel costs are high and environmental conditions harsh, especially if hydrogen is used not only to manage renewable energies but also to meet thermal loads and transport requirements.

Regarding the on-grid systems, hydrogen can be produced in large-scale centralized plants or at the end-use point. In centralized plants, the generation costs decrease due to the scale economy, electrolyzers are less expensive per kW installed and more efficient, and the capacity factor is higher depending on the management strategy. However, it is necessary to transport hydrogen to the end-use point and additional costs must be taken into account.

Hydrogen can be also produced at the end-use point, in filling stations for instance, avoiding transport costs. If wind power at these points is available at less than $0.038/kWh, it is possible for production, compression, and storage to cost below the target of $2–3/kg delivered hydrogen assessed by DOE according to the study developed by NREL in 2006 [30].

9.5 Conclusions

This study is a part of the activities carried out in the framework of the IEA Hydrogen Agreement, Task 24 "Wind Energy and Hydrogen Integration."

As a result of the study, we can conclude that hydrogen could compete with other ESSs, generally with batteries, mainly in energy applications of short and long term and linked to renewable energies such as wind power depending on the strategy applied. Hydrogen can be stored for a future reconversion into electricity or can be used in a different application, taking advantage of its energy vector feature.

Although there are still several disadvantages and technical problems to be solved such as low efficiency and high costs, the very high potential of improvement of the hydrogen technologies, and the numerous governmental programs supporting these research and developments, cause us to be optimistic concerning the future of hydrogen in the energy sector.

Most of the management strategies discussed here result in a rather low number of annual operating hours, leading to high specific costs and reducing the economic feasibility of wind-hydrogen systems. A major challenge will thus be to develop operating strategies that facilitate providing different services at different points in time (or even in parallel) to optimize the capacity factor and the costs per kilowatt-hour provided.

Nevertheless, it could be interesting to find new applications for hydrogen from wind power in addition to the energy storage with a higher added value as in a CHP application, or in the transport sector where most of the ESSs are not competitors and the rest are less efficient and/or versatile.

References

[1] Pure Power, Wind energy targets for 2020 and 2030. A report by the European Wind Energy Association—2009 update.

[2] http://www.20percentwind.org/20p.aspx?page=Report/.

[3] IEA World Energy Outlook, 2008.

[4] *Impacts of Large Amounts of Wind Power on Design and Operation of Power Systems; Results of IEA Collaboration.* Hannele Holttinen, Peter Meibom, Antje Orths, Mark O'Malley, Bart C. Ummels, John Olav Tande, Ana Estanqueiro, Emilio Gomez, J. Charles Smith, Erik Ela, Conference Paper, NREL/CP-500-43540 June 2008.

[5] K. Y. Cheung, S. T. Cheung, R. G. Navin De Silva, M. P. Juvonen, R. Singh, and J. J. Woo (2003). *Large-Scale Energy Storage Systems.* London: Imperial College London.

[6] A. Gonzalez, B. Ó Gallachóir, E. McKeogh, and K. Lynch (2004). *Study of Electricity Storage Technologies and Their Potential to Address Wind Energy Intermittency in Ireland.* Cork: University College Cork.

[7] Outlook of Energy Storage Technologies, IP/A/ITRE/ST/2007-07, February 2008.

[8] Electricity Storage Association, http://www.electricitystorage.org/site/technologies/.

[9] *Review of Energy Storage Requirements and Options, and the Benefits of Hydrogen,* Torgeir Nakken Statoil ASA.

[10] (a) http://www.hydrogensystems.be/. (b) http://www.hydro.com/electrolysers/en/. (c) http://www.teledyne.com/.

[11] (a) A. F. G. Smith and M. Newborough, "Low-Cost Polymer Electrolysers and Electrolyser Implementation Scenarios for Carbon Abatement," Edinburgh 2004. (b) L. S. Basye S. and S. Swaminathan, "Hydrogen Production Costs," U.S. Department of the Environment Report DOE/GO/10170-778, 1997.

[12] W. A. Amos, *Costs of Storing and Transporting Hydrogen*. Golden, Colorado, 1998.

[13] (a) http://www.fe.doe.gov/programs/powersystems/turbines/index.html. (b) M. P. Boyce, *Gas Turbine Engineering Handbook*, 2nd ed: Gulf Professional Publishing, 2002.

[14] (a) http://en.wikipedia.org/wiki/Reciprocating_engine. (b) http://www.eere. energy.gov/hydrogenandfuelcells/fuelcells/fc_types.html#phosphoric. (c) "Fuel Cells," The Institution of Electrical Engineers, 2003. (d) *Fuel Cell Handbook* (7th ed.), EG&G Technical Services, Inc. November 2004.

[15] (a) Susan M. Schoenung, *Characteristics and Technologies for Long- vs. Short-Term Energy Storage*, Sandia Report, SAND2001-0765, March 2001. (b) Susan M. Schoenung and William V. Hassenzahl, *Long- vs. Short-Term Energy Storage Technologies Analysis—A Life-Cycle Cost Study*, Sandia Report, SAND2003-2783, August 2003.

[16] H. Chen, T. N. Cong, W. Yang, C. Tan, Y. Li, and Y. Ding, Progress in electrical energy storage system: A critical review. *Progress in Natural Science* 2009; 19: 291–312.

[17] J. L. Bernal-Agustín and R. Dufo-López, Hourly energy management for grid-connected wind-hydrogen systems. *International Journal of Hydrogen Energy* 2008; 33(22):6401–6413.

[18] E. Troncoso and M. Newborough, Implementation and control of electrolysers to achieve high penetrations of renewable power. *International Journal of Hydrogen Energy* 2007; 32(13):2253–2268.

[19] C. Jørgensen and S. Ropenus, Production price of hydrogen from grid connected electrolysis in a power market with high wind penetration, *International Journal of Hydrogen Energy* 2008; 33(20):5335–5344.

[20] M. Aguado, E. Ayerbe, C. Azcárate, R. Blanco, R. Garde, F. Mallor, and D. M. Rivas, Economical assessment of a wind-hydrogen energy system using WindHyGen® software, *International Journal of Hydrogen Energy* 2009; 34: 2845–2854.

[21] S. R. Vosen and J. O. Keller, Hybrid energy storage systems for stand-alone electric power systems: optimization of system performance and cost through control strategies, *International Journal of Hydrogen Energy* 1999; 24: 1139–1156.

[22] S. Kélouwani, K. Agbossou, and R. Chahine, Model for energy conversión in renewable energy system with hydrogen storage, *Journal of Power Sources* 2004; 140: 392–399.

[23] A. G. Dutton, J. A. M. Bleijs, H. Dienhart, M. Falchetta, W. Hug, D. Prischich, and A. J. Ruddell, Experience in the design, sizing, economics, and implementation of autonomous wind-powered hydrogen production systems, *International Journal of Hydrogen Energy* 2000; 25: 705–722.

[24] L. N. Grimsmo, M. Korpaas, R. Gjengedal, and S. Moller-Holst, "A study of a stand-alone wind and hydrogen system," Nordic Wind Power Conference, 2004.

[25] E. Bilgen, Domestic hydrogen production using renewable energy, *Solar Energy*, 2004; 77: 47–55.

[26] R. Glöckner, Ø. Ulleberg, R. Hildrum, C. E. Grégoire, and P. Ife, Integrating Renewables for Remote Fuel Systems, *Energy & Environment* 2002; 13(4–5): 735–747.

[27] Ø. Ulleberg and T. L. Pryor, "Optimization of integrated renewable energy hydrogen systems in diesel engine mini-grids," WHEC 2002—14th World Hydrogen Energy Conference.

[28] (a) J. Burger and P. Lewis, "Energy storage for utilities via hydrogen systems," presented at 9th Intersociety Energy Conversion Engineering Conference, 1974. (b) R. Fernandes, "Hydrogen cycle peak-shaving for electric utilities," presented at 9th Intersociety Energy Conversion Engineering Conference, 1974.

[29] J. I. Levene, "An Economic Analysis of Hydrogen Production from Wind," NREL, Golden, CO NREL/CP-560-38210, May 2005.

[30] Set of the existing regulations and legislative framework related to RES implementation, June 2008.

[31] J. Levene, B. Kroposki, and G. Sverdrup, "Wind energy and production of hydrogen and electricity—Opportunities for renewable hydrogen," Conference Paper NREL/CP-560-39534, 2006.

10

Hydrogen Design Case Studies

Mathew Thomas, John W. Sheffield, and Vijay Mohan

CONTENTS

10.1 Introduction

Hydrogen technologies have the potential to increase energy security, stimulate economic growth, and reduce greenhouse gas emissions. There have been many projects worldwide that have proven the feasibility, durability, and reliability of hydrogen technologies. The original equipment manufacturers (OEMs) have conducted numerous demonstration programs around the world to document the lessons learned and establish best practice for safe operation of these technologies. The next step for these technologies would be the transition to commercial operation and the development of hydrogen infrastructure.

In 2009, California Fuel Cell Partnership (CaFCP) released an Action Plan[1] that outlines the strategies for the development of early "hydrogen communities" in the state of California to promote the transition of hydrogen technologies to real-world commercial operation. This chapter discusses real-world applications of hydrogen technologies for a hydrogen community. The applications are generic and are applicable for communities around the world. They include a commercial hydrogen fueling station, residential hydrogen fueling, hydrogen applications for airports, and other hydrogen applications These conceptual designs were created by the Missouri University of Science and Technology's hydrogen student design team in response to the Fuel Cell Hydrogen Energy Association (formerly known as National Hydrogen Association) Hydrogen Student Design Contests.

10.2 Commercial Hydrogen Fueling Station

The conceptual design of a commercial hydrogen fueling station includes a reproducible and scalable hydrogen fueling station for the city of Santa Monica, California. However, it should also be noted that the key elements of the design were selected such that the station can be replicated in other cities around the world. Emphasis was given to hydrogen produced from renewable sources such as biogas and landfill gas. At least 33% of hydrogen supplied at the station is produced from renewable sources to attract federal/state funding. Different options were evaluated during the design process, including the use of liquid hydrogen delivery, pipeline delivery, tube trailer delivery, on-site production, and so on. The final design includes on-site hydrogen production with tube trailer delivery of renewable hydrogen as shown in Table 10.1.

The station is capable of dispensing 400 kg of hydrogen per day and has the capacity to dispense hydrogen at both 5,000 and 10,000 psi. It is designed to dispense up to 240 kg of hydrogen continuously without suspending operations to recharge the storage tanks and have a high customer service level. Each dispenser can operate independently and can function even if one of the sources (fuel cell or tube trailer) is down for maintenance. The hydrogen station operation is discussed in detail in the later section of the chapter.

TABLE 10.1

Station Design Options

Technologies Evaluated during Design	Observations
Liquid delivery of renewable hydrogen	Expensive way to produce, store, transport hydrogen Losses from vaporization adds significant costs
Tube trailer delivery via two trailers	Daily delivery will be needed Connecting and disconnecting the trailer daily is cumbersome
On-site hydrogen production	Cannot provide 33% of renewable hydrogen
Pipeline	High infrastructure cost
On-site hydrogen production with tube trailer delivery of renewable hydrogen	Tube trailer rotation only at the end of third day, 33% renewable hydrogen supply, fuel cell eligible for 100% incentive

10.2.1 Hydrogen Production Evaluation

During the conceptual design process, facilities such as wastewater treatment plant, water reclamation plant, and landfill facility located within 25 miles of the city limits capable of producing of renewable hydrogen were identified. The wastewater treatment plant and water reclamation plant produce biogas (50%–80% methane)[2] during the water treatment process while the landfill facility generates landfill gas (50%–55% methane).[3] Methane present in these gases is separated to produce hydrogen via steam methane reformation using a combined heat, hydrogen, and power (CHHP) system. Hydrogen generated at these facilities is considered to be renewable hydrogen since it uses biogas and landfill gas as feedstock. Similar facilities producing biogas and landfill gas are also potential candidates for producing renewable hydrogen.

10.2.2 Hydrogen Compression and Transportation

Hydrogen produced at wastewater treatment plant and water reclamation plant is compressed on-site using a two-stage diaphragm compressor and transported to the hydrogen station via a hydrogen tube trailer. The two facilities have the capability to refill hydrogen tube trailers and the excess hydrogen can be sold commercially. The landfill facility has a hydrogen bottling plant capable of refilling hydrogen cylinders, which can be used by the early fuel cell market customers in the community. It also has the capability to refill the mobile hydrogen refueler, which can be used to fill the material handling vehicles and other hydrogen-powered vehicles. Figure 10.1 illustrates different operations at the three facilities including hydrogen production, compression, storage, and transportation. The utilization of compressed hydrogen is discussed in the following sections.

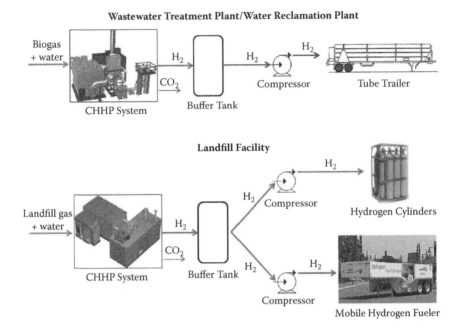

FIGURE 10.1
Hydrogen production, compression, and transportation. (From Mohan, V., Design of a Hydrogen Community for Santa Monica [thesis], Missouri University of Science and Technology, 2011. http://scholarsmine.mst.edu/thesis/pdf/Mohan_09007dcc80917065.pdf.)

10.2.3 Fueling Station Location and Site Plot

A location was selected based on its proximity to current hydrogen refueling stations, potential customers, as well as access and visibility from freeways through the city. The rectangular lot would accommodate the refueling area for vehicles, maximum hydrogen storage capabilities, and a sustainably designed convenience store. The station refueling area has two hydrogen pumps available to either the left or right side of a vehicle. The vehicle refueling area, as well as the gated entrance next to the hydrogen storage area, provides highly visible shutoff buttons in the case of an emergency. A two-hour firewall surrounds the hydrogen storage area, shielding the convenience store and local sidewalks in the event of failure. There is no roof over the hydrogen storage area allowing ample ventilation. The electrical control panel for the hydrogen area is located within the convenience store, keeping it safe and easily accessible by any clerk or attendant. As a precaution, a flame sensor system as well as security cameras were placed at strategic locations to monitor activities within the hydrogen storage area. The design of the hydrogen storage area was located so that a hydrogen delivery

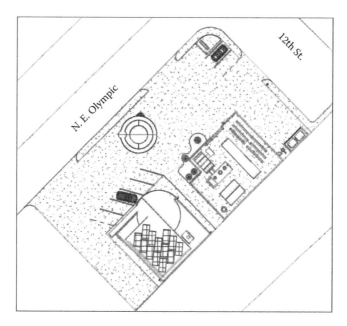

FIGURE 10.2
Plot layout of commercial hydrogen fueling station. (From Mohan, V., Design of a Hydrogen Community for Santa Monica [thesis], Missouri University of Science and Technology, 2011. http://scholarsmine.mst.edu/thesis/pdf/Mohan_09007dcc80917065.pdf.)

truck would be able to easily place a trailer within the equipment area for the hydrogen delivery.

The site plot and the different views of the hydrogen station and convenience store can be found in Figures 10.2 and 10.3.

10.2.4 Hydrogen Fueling Station Components

Solar and wind energy in the region were found not to be favorable for on-site production of renewable hydrogen. As mentioned earlier, the target is to supply at least 33% renewable hydrogen at the station. Hence, two hydrogen sources—(1) renewable hydrogen (produced off-site from biogas and landfill gas) and (2) nonrenewable hydrogen (produced on-site from natural gas)—were chosen to supply hydrogen at the station. Renewable hydrogen produced off-site is transported to the station via tube trailer. The tube trailer is capable of carrying 300 kg of hydrogen at 3,000 psi and is exchanged for a new one every three days to meet the 33% renewable hydrogen criteria. The on-site hydrogen production is through a molten carbonate fuel cell capable of generating 136 kg/day hydrogen and 250 kW of electric power using steam

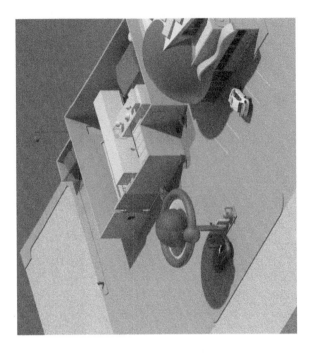

FIGURE 10.3
Commercial hydrogen fueling station and convenience store. (From Mohan, V., Design of a Hydrogen Community for Santa Monica [thesis], Missouri University of Science and Technology, 2011. http://scholarsmine.mst.edu/thesis/pdf/Mohan_09007dcc80917065.pdf.)

methane reformation of natural gas. The station has a total hydrogen storage capacity of 198 kg (132 kg at 5,000 psi and 66 kg at 10,000 psi) and has two dispensers capable of dispensing hydrogen at both 5,000 psi and 10,000 psi. The maximum allowable time for fueling the vehicles is 5 minutes for the 5,000 psi vehicle and 7 minutes for the 10,000 psi vehicle. The system is designed such that both dispensing units can work independently of each other. Figure 10.4 shows the major components of the hydrogen delivery, production, compression, and storage system.

Table 10.2 lists the major components of the fueling station and its specifications.

The station also has safety equipment, including hydrogen and flame detection systems, safety cameras, hydrogen storage isolation systems, and pressure relief devices to prevent and/or mitigate possible dangerous events at the fueling station. Safety barriers and walls will protect the dispenser and hydrogen equipment at the station. Emergency shutdown switches have been strategically located at the fueling station and will terminate all operations at the station when activated. Other safety features included breakaway hoses, fire extinguishers, and safety signs.

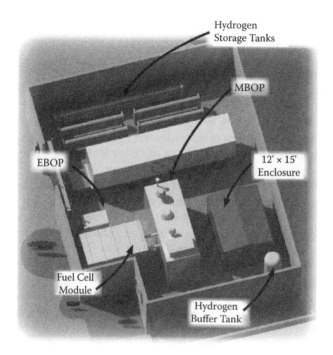

FIGURE 10.4
Hydrogen production, compression, and storage area. (From Mohan, V., Design of a Hydrogen Community for Santa Monica [thesis], Missouri University of Science and Technology, 2011. http://scholarsmine.mst.edu/thesis/pdf/Mohan_09007dcc80917065.pdf.)

TABLE 10.2

Specifications of Major Components

#	Item	Specification
1	Fuel cell	250 kW/136 kg H_2/day
2	Buffer tank	~ 5 kg @ 200 psi
3	Tube trailer	300 kg @ 3,000 psi
4	Hydrogen compressor	120 psi–3,000 psi @ 40 scfm
5	Hydrogen compressor	120 psi–6,000 psi @ 90 scfm
6	Hydrogen compressor	6,000 psi–12,000 psi @ 130 scfm
7	Storage unit (5,000 psi)	132 kg
8	Storage unit (10,000 psi)	66 kg
9	Hydrogen dispenser	5,000 psi & 10,000 psi

10.3 Residential Fueling with Hydrogen

Even though major automotive companies and energy companies have invested heavily into the hydrogen transportation sector, transition to using hydrogen as a fuel has been very slow. A major obstacle to this transition is the lack of hydrogen fueling infrastructure. An innovative solution to this problem is the development of a residential hydrogen fueling system. With residential fueling, the customers have flexible refueling options and do not have to rely entirely on commercial hydrogen fueling stations.

Residential hydrogen fueling portrayed in this chapter uses solar photovoltaic (PV) panels and produces renewable hydrogen through electrolysis. A total of sixty 315 W solar panels are installed across the roof and the power produced from them is used to generate renewable hydrogen through electrolysis. The electrolyzer and dispenser are housed in an insulated portion of the garage whereas the hydrogen is stored in a separate enclosure outside the garage (Figure 10.5). The building's design incorporates a ground source heating and cooling system to efficiently keep the interior of the home at a constant, comfortable temperature throughout the year. To protect the electrolyzer against freezing temperatures, a heat loop will feed back to the electrolyzer, keeping it at a constant temperature of 50–55°F during the winter months. The vestibule located in the front of the house creates an "air lock" entry, minimizing the loss of conditioned air as people enter and leave the Hydrogen Home.

All electrical equipment and outlets inside the garage are installed in accordance with the safety codes and standards. A hydrogen detector is also installed at the highest part of the roof. Roof vents are located on the very top of the

FIGURE 10.5
Hydrogen system at Hydrogen Home. (From Shah, A., V. Mohan, J. W. Sheffield, and K. B. Martin, *International Journal of Hydrogen Energy*, 36(20), 2011, 13132–13137.)

garage, providing ample ventilation. A mechanical ventilation system is also provided and will turn on when hydrogen concentration within the garage is more than 25% of the lower flammability limit. There are no solar panels over the hydrogen equipment so that the expensive solar panel equipment will not be damaged in the extreme case of failure. The concrete-insulation-concrete (C-I-C) walls utilized in the design are rated for two-hour fire resistance and a firebreak wall was incorporated into the garage to protect the house in the event of a failure of the equipment. The hydrogen dispenser, as well as the outdoor hydrogen storage enclosure, is surrounded with concrete car barriers to protect against accidental collisions into the equipment. An AC power disconnect is included in the design to isolate solar panels and inverters from the grid.

10.3.1 Vehicle Parameters

The residential hydrogen fueling system was designed primarily for hydrogen vehicles that follow noncommunication fill protocol and is able to partially fill vehicles with 5,000 psi or 10,000 psi storage tanks. The hydrogen vehicle requirements assumed during the design are shown in Table 10.3. Cascading results and the amount of hydrogen transferred into the vehicle are discussed in Section 10.3.3. For the design, it was assumed the total onboard storage capacity of the vehicle is 4.2 kg of gaseous hydrogen at 5,000 psi.

10.3.2 Hydrogen Fueling System Components

Various hydrogen production alternatives were investigated during the design to meet the on-site hydrogen demand and are discussed below.

1. Steam-methane reformation—Though it has an efficiency of up to 80%, it was not selected due to the following reasons:
   a. Daily hydrogen requirement is low.
   b. High capital cost and low efficiency of small steam methane reformation system. Capital cost was estimated to be $750/kW for 1 million scf/day (2,544 kg/day) and $4,000/kW for 100,000 scf/day (254 kg/day) steam methane reformation system respectively.[4] This is only expected to increase as the capacity decreases.

TABLE 10.3

Hydrogen Vehicle Requirements

Parameter	Value
Annual mileage	12,000 miles
Daily commute	35 miles total
Fuel economy	44 miles per kilogram
Daily requirement	0.8 kilogram of hydrogen

c. The reaction results in the emissions of carbon dioxide, which is a greenhouse gas.

d. A very high temperature (of 1000°C) is required to support steam-methane reaction.

e. Price of natural gas is highly fluctuating; thus the cost of hydrogen produced is very unreliable.

2. Reversible proton exchange membrane (PEM) fuel cell—Due to the inefficiency associated with the reversible fuel cells, this method is only economical if hydrogen is produced during off-peak periods or if electricity is produced during peak periods. It was not chosen as there are currently no commercially available reversible PEM fuel cells capable of producing at least 1 kg/day of hydrogen.

3. Low-temperature electrolysis—This method is preferred due of the following reasons:

a. There are no carbon dioxide emissions if the electrolyzer is coupled with renewable energy sources.

b. A fairly low temperature (<100°C) is involved.

c. Cost of hydrogen depends more on electricity price, which is fairly stable.

d. Capital cost is estimated to be $2,000/kW–$4,000/kW for 3–12 kg/day.[5] This is expected to decrease with technological advancements.

Based on the daily hydrogen requirement (0.8 kg/day), a high-pressure hydrogen generator (Figure 10.7) was chosen since it has the low energy consumption, high efficiency, and high output pressure. The hydrogen generator can produce up to 2.28 kg of hydrogen per day at 2,400 psi and can be located indoors in a nonhazardous/unclassified area. The use of a high pressure electrolyzer also negates the need for a hydrogen compressor, which is a major source of energy loss.

10.3.3 Hydrogen Fueling Operation

The high-pressure hydrogen produced on-site is stored in three hydrogen storage tanks. Figure 10.8 shows the major equipment used for residential hydrogen fueling.

The set point of the electrolyzer is adjusted such that the unit will turn on if the pressure in any of the three tanks falls below 2,000 psi and will turn off when all three storage tanks reach 2,400 psi. A hydrogen priority panel is used to prioritize filling of the three hydrogen storage tanks and sequence the flow of hydrogen from the storage tanks to the dispenser. This system is designed such that the maximum allowable ramp rate (MARR) is not exceeded during hydrogen dispensing. Pressure transducer inside the

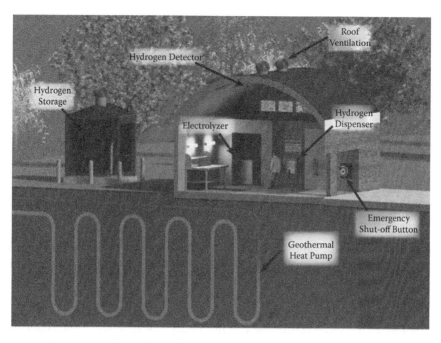

FIGURE 10.6
Major components of the residential hydrogen fueling station.

FIGURE 10.7
High-pressure hydrogen generator. (From Shah, A., V. Mohan, J. W. Sheffield, and K. B. Martin, *International Journal of Hydrogen Energy*, 36(20), 2011, 13132–13137.)

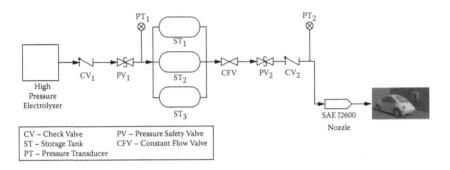

FIGURE 10.8
Detailed report of cascade simulation. (From Shah, A., V. Mohan, J. W. Sheffield, and K. B. Martin, *International Journal of Hydrogen Energy*, 36(20), 2011, 13132–13137.)

dispenser determines the pressure inside the onboard vehicle storage tank and vehicle fueling is initiated only if it is less than the pressure inside the hydrogen storage tanks. A nozzle complaint with SAE J2600 standard is used for dispensing the hydrogen into the vehicle.

Since the design employs a high-pressure electrolyzer with an output of 2,400 psi, a hydrogen compressor is not necessary to achieve the daily fill requirement. The amount of hydrogen transferred from the storage tanks into the fuel cell vehicle can be estimated using CASCADE™ gaseous fueling system sizing software.[6] Figure 10.9 shows the detailed report of the simulation using the CASCADE™ software. In the simulation, the fuel cell vehicle is initially at 502 psi and has 0.54 kg (5,975 liters at STP) of hydrogen

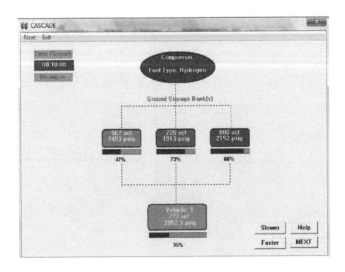

FIGURE 10.9
Hydrogen applications for airport. (From Shah, A., V. Mohan, J. W. Sheffield, and K. B. Martin, *International Journal of Hydrogen Energy*, 36(20), 2011, 13132–13137.)

stored on board. At the end of the fueling event, the hydrogen storage on board the vehicle is at 2,052 psi and contains 1.97 kg (21,889 liters at STP) of hydrogen. Hence, a total of 1.43 kg of hydrogen is transferred during the fueling operation and is more than the daily requirement of 0.8 kg. Also, from these results, it can be seen that the vehicle can be filled in 10 minutes without the use of a compressor. This both saves capital costs and reduces the operational cost of the residential hydrogen refueling. If higher hydrogen transfer is required, a hydrogen compressor can be added to the system and higher transfer quantity can be achieved. But it should be noted that the current hydrogen fueling system has few mechanical parts and will have higher reliability than a system using a compressor.

It is advised that the fueling operation be performed after the vehicle reaches equilibrium with the ambient temperature in the garage. Since the hydrogen fueling requirement per day can be achieved using the selected fueling system, a prepackaged or commercial hydrogen 5,000 psi dispenser is not required. This results in significant cost reduction for the hydrogen residential fueling system. The car can be fully refueled from a commercial station for occasional long distance travel of more than 35 miles.

Safety was the most important factor during the design of the residential hydrogen fueling system. Safety equipment in the garage include hydrogen detector, breakaway hoses, easily accessible emergency shutdown buttons, fire extinguishers, hydrogen storage isolation system, and pressure relief devices. Safety barriers were included in the design to protect the dispenser and hydrogen equipment. Emergency lights, alarm, and a mechanical ventilation fan rated for hazardous locations (Class 1 Div 2, Group B) will turn on in case of hydrogen leak detection by the gas detector. Electrical power to the rest of the garage will be disconnected to prevent ignition of hydrogen from electrical sources.

10.4 Hydrogen Applications for Airports

Airports face unique challenges related to air and water quality, noise pollution, energy efficiency, and safety and security. Furthermore, airports around the world are being faced with the difficulty of operating in an economically sound way. As problems such as pollution, flight delays, and security risks escalate, airports will find themselves in an even more challenging situation. Hydrogen and hydrogen technologies have the potential to mitigate these problems. Hydrogen applications selected in the design include a hydrogen fueling station, auxiliary power generation, energy savings power generation, hydrogen fuel cell vehicles/remote devices, and technologies dedicated to public education. To facilitate a better systems understanding, the applications have been represented visually in Figure 10.10.

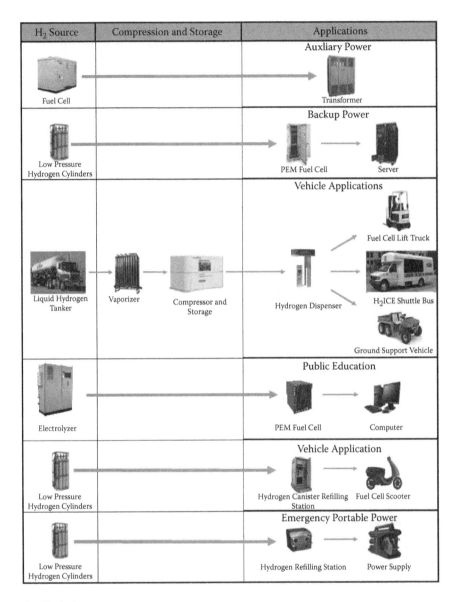

FIGURE 10.10
Hydrogen applications for airports.

10.4.1 Hydrogen Fueling Station

To reduce initial capital costs and to maintain a flexible, evolving hydrogen system, the design uses liquid hydrogen purchased from an external source. Liquid hydrogen from a hydrogen tanker is transferred to a cryogenic tank on-site and is converted into gas using a vaporizer. Gaseous hydrogen from the vaporizer can be compressed and stored at high pressure or can be fed directly to the hydrogen fueling station. The system has a separate dispenser and is capable of dispensing gaseous hydrogen at 5,000 psi.

10.4.2 Auxiliary Electric Power Generation

During the analysis of flight operations, it was observed that airports are experiencing an increase in power outages. This negatively impacts airport operations, and hence the design includes an auxiliary power system to address this critical weakness. The backup power system has a capacity of 30 kW made possible through the stacking of six 5 kW fuel cell modules. They are equipped with fuel leak sensors and remote communication and control ability. An additional module that includes batteries is used to bridge the downtime between a power failure and fuel cell warm-up time to achieve backup power functionality. While this backup power could serve any number of different areas, this design suggests that the airport computer network be protected first. The reliability of the fuel cells and the backup power unit in general will ensure that no data is lost in the event of a power outage. By utilizing this system, airports will experience fewer critical outages, ultimately not only preserving their flight schedules but also reducing the effects of outages throughout the country.

10.4.3 Energy Savings Electric Power Generation

Additional power generation technologies have been selected to reduce on-peak loading by approximately 200 kW and to drive down energy costs. The fuel cell power plant (Figure 10.11, selected during the design, reforms natural gas to generate electric power. The module delivers 200 kW of electrical power as well as 900,000 Btu/hr of heating capability. This heat recovery is possible through a double wall heat exchanger, allowing potable water to be processed. The system is also capable of processing landfill gases. This ability presents opportunities for future project work to further improve environmental impacts. The system can be configured to run with either 400 V at 50 Hz or 480 V at 60 Hz.

The footprint of the power module is 9'6" × 18', allowing a single unit to be installed in a variety of locations. During emergency situations, this equipment also acts as auxiliary power generation. Of the many advantages this unit offers, perhaps the most notable is that it is capable of running for long periods of time as long as fuel is readily available. With only a single unit, per unit specifications power assurance is in excess of 99.99%.

10.4.4 Hydrogen Fuel Cell Vehicles/Remote Devices

The design includes the use of different hydrogen technologies to address several personal vehicle and remote power applications. These vehicle applications are a hydrogen internal combustion engine (H_2ICE) shuttle bus, a

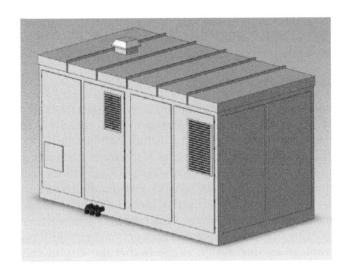

FIGURE 10.11
Rendering of the fuel all system. (From Shah, A., V. Mohan, J. W. Sheffield, and K. B. Martin, *International Journal of Hydrogen Energy*, 36 (20), 2011, 13132–13137.)

hydrogen fuel cell lift truck, a hydrogen fuel cell ground support vehicle, and hydrogen fuel cell scooters, as well as hydrogen personal power packs.

The most noticeable hydrogen vehicle included in the design is the shuttle bus (Figure 10.12). This vehicle transports passengers from the airport to nearby hotels, parking lots, and other points of interest.

To meet the needs specified for a hydrogen fuel cell lift truck, a fuel cell power pack capable of delivering 21 kWh of electricity at 48 V was selected. The power pack shown in Figure 10.13 can be used to power a standard class 1 lift truck. The advantages of this technology are reduced fueling times, elimination of space for charging stations, extended run-time between fills, and

FIGURE 10.12
Hydrogen internal combustion engine shuttle bus.

FIGURE 10.13
Fuel cell power pack.

an emissions-free power source. This is especially useful if the equipment is being used inside where ventilation is less than adequate. Additionally, this fuel cell module has been designed such that they are a drop-in replacement for traditional battery power systems.

The fuel cell power pack can also be used with the ground support vehicle. This vehicle can be used in terminal for light cargo as well as passenger transport. If necessary, the vehicle could also be used for other applications throughout the airport, both inside and outside any structure. Like the fuel cell lift trucks mentioned above, this vehicle can also operate emissions free and has a reduced refueling time when compared to existing systems.

To serve as an additional roaming advertisement to today's hydrogen possibilities, the design incorporated the use of a hydrogen fuel cell scooter (Figure 10.14). The hydrogen fuel cell scooter boasts a power plant producing

FIGURE 10.14
Fuel cell scooter.

FIGURE 10.15
Portable fuel cell.

120 amps at 24 V and can reach a maximum level speed of just over 30 mph. At a more moderate speed of 18 mph, the scooter has a range of approximately 37 miles before refueling is necessary. The scooter's fuel supply is delivered via a metal hydride canister that is simply exchanged for a new canister at refueling. The scooter and fuel canister have a combined weight of 240 pounds, allowing the vehicle to operate nearly anywhere pedestrian traffic is possible.

Finally, the design includes fuel cell power units (Figure 10.15) for remote power applications such as communications equipment for first responder teams, small electric tool operation, or any other application that requires light, reliable portable electric power. Its advantages over conventional battery units are a more compact size, a consistent run-time that does not degrade with age, and no self-discharge, giving the unit a very long shelf life.

10.4.5 Public Education Technologies

The last application for which hydrogen technologies have been selected for use is that of a public/passenger education center. To ensure optimal public exposure, the decision was made to place it in the public area of the airport, near the ticketing desks. This center, located in a high-traffic area of the airport, demonstrates where hydrogen is being used within the airport and the greater possibilities that can be realized through the use of hydrogen technologies. The center is powered entirely by hydrogen produced through electrolysis powered by solar panels. The electrolyzer produces 4.31 kg of hydrogen per day, which in turn fuels a 5 kW hydrogen fuel cell powering multiple computers, as well as audio/visual equipment located within the center.

The technologies selected for this design should not be seen as the end product of a hydrogen infrastructure at an airport. Instead, these systems have been designed to serve as a stepping-stone to the introduction of larger hydrogen systems within an airport or similar facility.

10.5 Additional Hydrogen Applications for the Community

Different areas in the community that could potentially use hydrogen technologies were identified. They include facilities such as schools, banks, offices, hospitals, warehouses, and parks. Hydrogen applications for these facilities include backup power, material handling, public transport, personal transport, portable power, and public education centers.

10.5.1 Backup Power

Hydrogen powered fuel cells can be used as a source of backup power at different locations including schools, universities, banks, hospitals, and communication towers. Currently, most of the backup power equipment is powered by batteries and generators, which are mechanically driven and hence incur a high maintenance cost. Fuel cells have high reliability and durability along with lower maintenance cost and noise pollution. A 5 kW PEM fuel cell (Figure 10.16) was selected to support the different customers and their critical operations. These fuel cell modules can be used in modular configuration to provide up to 30 kW of electric power. Sensors for fuel leak detection are included and these are equipped with remote communication. Table 10.4 shows the potential customers identified for providing backup power.

FIGURE 10.16
5 kW fuel cell module.

TABLE 10.4

Potential Customers Identified for Providing the Backup Power

Customer	Fuel Cell Used	Use of Fuel Cells	Hydrogen Source
Schools	Five stacks of 5 kW fuel cells generating 25 kW electric power	Powers UPS and computers	Hydrogen cylinders delivered on-site
Banks	Five stacks of 5 kW fuel cells generating 25 kW electric power	Backup power for computers, servers, alarms, communication lines, security cameras, token generator, etc.	Hydrogen cylinders delivered on-site
Hospitals	Fuel cell generating 300 kW of electric power using reformation of natural gas	Powers emergency lights, equipment in operation theater, critical units, elevators, surveillance cameras, etc.	Natural gas reformed to generate hydrogen on-site
State parks	Two stacks of 5 kW fuel cells generating 10 kW electric power	Raise public awareness about hydrogen technologies, provides power to the building, demonstration of fuel cell	Hydrogen cylinders delivered on-site

10.5.2 Material Handling

Lift trucks, utility vehicles, and the other material handling equipment are used extensively in large manufacturing facilities, warehouses, and stores for transporting heavy material at increased operational efficiency. It is recommended that these facilities replace the current battery and internal combustion engine powered forklifts with fuel cell–powered forklifts. PEM fuel cells are more cost effective than batteries, which have to be constantly replaced while running two or three shifts per day. Table 10.5 shows the potential customers identified for providing fuel cell material handling equipment.

TABLE 10.5

Potential Customers for Material Handling Equipment

Customer	Use	Hydrogen Source
Warehouses	Transportation of hardware	On-site fueling using mobile hydrogen fueler
City / Utility Company	Transportation of heavy materials	On-site fueling using mobile hydrogen fueler
Stores	Transportation of goods	On-site fueling using mobile hydrogen fueler

FIGURE 10.17
Fuel cell transit bus.

10.5.3 Portable Power

Fuel cells also can be used in portable power applications such as law enforcement, surveillance/security cameras, emergency response, and telecommunication purposes. Fuel cells are extremely energy efficient and have significantly longer run-times than batteries and also provide a clean and a reliable way of generating portable power. The portable fuel cells currently available in the market can supply peak power of 330 W and have run-times of up to 2160 Wh.

10.5.4 Public and Private Transportation

H_2ICE or fuel cell transit buses (Figure 10.17) can be used in cities having high air pollution. By operating in highly congested areas, these buses can reduce pollution and traffic problems. Fuel cell vehicles can also be used as fleet vehicles by companies or corporations and by customers for personal use. Fuel cell vehicles do not have any tail pipe emissions and are completely emission-free if the hydrogen is produced from renewable energy sources.

10.5.5 Public Education Center

The Hydrogen Education Center is designed from six recycled cargo containers and conventional building materials with concrete masonry unit walls. It has an open lobby, a large educational classroom, men's and women's restrooms, a green roof observation deck, and a multipurpose space on the second floor. In addition to being environmentally friendly, the Hydrogen Education Center also includes space capacity for eight large solar panels. The ideal location for this conceptual building would be at the entrance grounds to a park or near a bike path. Due to the modular nature of these containers, it can be easily transported to other locations if necessary. Figure 10.18 shows a rendering of this facility.

FIGURE 10.18
Public education center.

10.6 Conclusion

The chapter discusses different hydrogen technology applications that can be incorporated into a community. The hydrogen technologies used in this design should be seen as a stepping-stone toward developing a hydrogen community and should include more hydrogen technologies as it evolves. They can be used separately or collectively in different areas of the community such as schools, hospitals, banks, offices, parks, airports, transportation, warehouses, and stores.

10.7 Acknowledgments

The authors wish to thank the Hydrogen Design Solutions Team at the Missouri University of Science and Technology.

References

1. *Hydrogen Fuel Cell Vehicle and Station Deployment Plan: A Strategy for Meeting the Challenge Ahead*, Action Plan, California Fuel Cell Partnership, February 2009, http://www.cafcp.org/sites/files/Action%20Plan%20FINAL.pdf.

2. Alternative Fuels and Advanced Vehicles Data Center, U.S. Department of Energy, *What Is Biogas?*, http://www.afdc.energy.gov/afdc/fuels/emerging_biogas_what_is.html.
3. Landfill Methane Outreach Program, U.S. Environmental Protection Agency, http://epa.gov/lmop/faq/landfill-gas.html.
4. Ogden, J. *Review of Small Stationary Reformers for Hydrogen Production.* International Energy Agency, Report IEA/H2/TR-02/002, 2002.
5. Naterer, G.F.; Fowler, M.; Cotton, J.; Gabriel, K. *Synergistic Roles of Off-Peak Electrolysis and Thermochemical Production of Hydrogen from Nuclear Energy in Canada*, 2008.
6. CASCADE™ Gaseous Fueling System Sizing Software, Gas Technology Institute and InterEnergy Software.

Section III

Hydrogen Safety

11

Hydrogen Safety

Mathew Thomas

CONTENTS

11.1 Introduction

Hydrogen has the potential to help provide clean energy for a variety of uses including stationary, portable, and automotive energy applications. Hydrogen can provide clean and secure energy in a future of increasing energy demands and rapidly depleting fossil fuels and can be produced from a wide range of sources including natural gas, water, and biogas. Although hydrogen has been used for decades by the National Aeronautics and Space Administration (NASA) and in various industrial applications including production of chemicals, metals, electronics, and petroleum, it is perceived to be a dangerous fuel. This chapter looks at hydrogen safety in a hydrogen infrastructure setting such as hydrogen fueling stations, hydrogen vehicle research and development garages, hydrogen storage, and stationary fuel cell installations. Before moving into these topics, it is important to understand the properties of hydrogen that differentiates it from other gases and fuels.

11.2 Unique Properties and Hazards

Hydrogen is the most abundant element in the universe, but it is relatively rare in its natural form and occurs as compounds such as water and hydrocarbons. It is colorless, odorless, and tasteless, making it hard to detect. To safely produce, handle, and use hydrogen it is essential that one be familiar with the basic properties of hydrogen and receive basic hydrogen safety training. A summary of the properties of hydrogen is given in Table 11.1.

Due to the unique properties of hydrogen, there are different hazards associated with its production, use, and handling. They include the following:

- Flammability
- High pressure and over-pressurization
- Asphyxiation
- Nearly invisible flame

TABLE 11.1

Properties of Hydrogen

Properties	Value
Density (NTP)[a]	0.08375 kg/m^3
Viscosity (NTP)[a]	8.81 x 10^{-5} g/cm-sec
Specific gravity (NTP) (air = 1)[b]	0.0696
Normal boiling point[a]	−252.9°C
Auto ignition temperature in air[c]	585°C
Flammability limits in air[c]	4.0–75.0%
Flame temperature in air[c]	2045°C
Internal energy (NTP)[a]	2648.3 kJ/kg
Ignition energy[b]	0.02 mJ

[a] NIST Chemistry WebBook, http://webbook.nist.gov/chemistry/.
[b] Hydrogen Fuel Cell Engines and Related Technologies. Module 1: Hydrogen Properties, U.S. Department of Energy. 2001, http://www1.eere.energy.gov/hydrogenandfuel-cells/tech_validation/pdfs/fcm01r0.pdf.
[c] National Aeronautical and Space Administration, Safety Standard for Hydrogen and Hydrogen Systems (NSS 1740.16), http://www.hq.nasa.gov/office/codeq/doctree/canceled/871916.pdf.

11.2.1 Flammability

The accumulation of hydrogen is a cause for specific concern, since hydrogen (like other flammable gases) has a tendency to detonate when ignited in confined areas. The wide flammability limits (4.0%–75.0% in air) and low minimum ignition energy (0.02 mJ) exacerbate this issue. There are three distinct aspects to the mitigation of this hazard: sensors for early detection, ventilation to prevent accumulation, and electrical grounding to avoid static discharges. Sensors are necessary because hydrogen is naturally an odorless gas and odorants that are typically added to gases (such as the methanethiol added to the natural gas used in homes) cannot be used with fuel cells, as they degrade the catalyst. The high buoyancy of hydrogen potentially makes ventilation simpler than it would be for other fuels. An open vent at the highest point could well be sufficient to evacuate hydrogen to the outside environment. It is also recommended that hydrogen be stored outdoors or space with ample ventilation, allowing rapid diffusion of hydrogen in case of a leak. To avoid static charges as an ignition source, fuel cells and other equipment should be electrically grounded. This requires proper installation and maintenance of the building's wiring system. Hydrogen systems must be designed in such a way that the hydrogen does not come in contact with the potential sources of ignition including electrical, mechanical, thermal, and chemical sources. Since hydrogen flame is almost invisible during the day and has very low radiant heat, flame detection systems should also be installed at facilities using hydrogen.

11.2.2 High Pressure and Over-pressurization

Since hydrogen has very low density, it is either liquefied or compressed to high pressures to store an adequate amount of hydrogen. Generally, hydrogen is compressed up to 2,400 psi for laboratory applications and up to 10,000 psi for hydrogen vehicle application. Mechanical failure of storage tanks, piping, compressors, and other equipment can cause explosions or physical damage. Liquid hydrogen expands to more than 800 times its volume when it vaporizes into gaseous state, which could lead to over-pressurization and mechanical failure of the system. Therefore, high-pressure hydrogen systems must be equipped with pressure relief devices or valves to prevent over-pressurization. Proper personal protective equipment (PPE) must be worn while working with high-pressure hydrogen.

11.2.3 Asphyxiation

Hydrogen, being colorless, odorless, tasteless, and nonirritating, cannot be detected by human senses and can cause asphyxiation in confined spaces at high concentrations. Liquid hydrogen spills can cause asphyxiation and can damage the lungs. This can be avoided by providing adequate ventilation and installing hydrogen detectors.

11.2.4 Nearly Invisible Flame

Hydrogen flame is different from flames of hydrocarbon fuels and burns with a pale blue flame, which is very nearly invisible under normal daylight conditions. Besides low visibility, hydrogen flames have low radiant heat, which diminishes the ability of a person to detect a hydrogen fire. Hence, a hydrogen flame may not be felt on the skin until a person is within the flame itself.

Due to the unique properties of hydrogen and the hazards associated with it, facilities involving hydrogen must follow safety protocols for its safe handling. Several organizations, including the National Fire Protection Association (NFPA), the International Code Council (ICC), the American National Standards Institute (ANSI), Underwriters Laboratories Inc. (UL), the Compressed Gas Association (CGA), the Society of Automotive Engineers (SAE), and American Society of Mechanical Engineers (ASME) have developed codes and standards for the safe handling and use of hydrogen and hydrogen technologies. Standards are available covering a wide range of possible safety issues. The following section briefly discusses the safety features that these facilities must have to operate in a safe manner.

11.3 Hydrogen Fueling Station

Hydrogen fueling stations can be located either indoors or outdoors in compliance with NFPA 2[1] and NFPA 52.[2] Outdoor fueling stations are usually used to fill hydrogen-powered vehicles (e.g., fuel cell cars, buses, hydrogen internal combustion engine vehicles), whereas the indoor fueling stations are primarily used to fill hydrogen-powered industrial trucks or forklifts. In both cases, all hydrogen equipment must be located inside a protected area to protect from physical damage or vandalism.[1,2] As additional safety features, both outdoor and indoor facilities must have fire extinguishers, safety bollards or barriers protecting the dispenser from vehicle impact, and breakaway devices for dispenser hoses in the event of a drive-off.[1,2] As per regulations, proper warning signs including "WARNING—NO SMOKING—FLAMMABLE GAS," "NON-ODORZIED GAS," and "NO CELL PHONE" must be placed at the station.[1,2] Codes and standards applicable for hydrogen fueling stations are tabulated in Table 11.2.

Figure 11.1 shows the minimum distance of a 5,000 psi outdoor gaseous hydrogen system to exposures in accordance with NFPA 52.[2] The minimum distances to exposures are established based on the pressure of the hydrogen system and the maximum internal diameter of interconnecting piping. The setoff distances will change based on the operating pressure and internal diameter of piping.

For safety reasons, hydrogen fueling should be done only by a qualified operator. Often, a PIN or an access card is required to operate the dispenser.

TABLE 11.2

Codes and Standards for Hydrogen Fueling Station

Category	Codes and Standards
Fueling Station Design	U.S. Department of Energy—Hydrogen Fueling Station Codes and Standards, International Fire Code, International Fuel Gas Code, NFPA 55, NFPA 52
Dispensing Equipment	SAE J2600, NFPA 52, ISO 17268, SAE J2783, SAE J2601, SAE J2799—TIR, CSA America HGV4, USA National Institute of Standards and Technology (NIST) Weights and Measures Division, OILM R 81, OILM R 139
Installation	U.S. Department of Energy—Hydrogen and Fuel Cells Permitting Guide, Handbook for Approval of Hydrogen Refueling Stations
Tanks & Storage	ASME Boiler & Pressure Vessel Code, CGA Publication H3, CGA Publication PS17, CGA Publication PS20, CGA Publication PS21, CGA Publication PS25, CGA Publication PS26, CGA Publication PS33, NFPA 55, ASME Boiler and Pressure Vessel Code Section XII- Transportation Tanks, CGA Publication H6
Piping & Pipelines	ASME B31 Series, CGA Publication G5.4, CGA Publication G5.6, ASME B31.12, CGA Publication G5.8, CGA Publication G5.7 (EIGA Doc 120/04)
Venting	CGA Publication G5.5
Labeling	CGA Publication H2
Safety/General Design	CSA International Requirement No. 5.99, Outline of Investigation UL Subject 2264B, ISO 16110-1, UL Subject 2264A, UL Subject 2264 C (Joint activity with CSA America; FC5), ISO TC197 Working Group 8/ISO 22734-1:2008
Safety	US Department of Labor, OSHA: 29 CFR 1910.103, AIAA G-095, CGA Publication P12, ISO TR 15916, European Integrated Hydrogen Project (EIHP)—Work Package 5

Source: U.S. Department of Energy, Hydrogen and Fuel Cells Program, http://www.hydrogen.energy.gov/permitting/fueling_stations.cfm.

This will restrict access to the hydrogen dispenser and will make sure that unauthorized fueling is not performed. Before beginning the fueling operation, the station and the vehicle should be electrically bonded to avoid electric discharge and potential fire or explosion. The dispenser should be equipped with fire and gas detectors to detect fire and hydrogen leaks in the system. The detectors should be connected to audible and visible alarms that will be activated in the event of a fire or hydrogen leak (Figure 11.2).

A manual emergency shutdown device (ESD) must be located near the dispenser and other major locations of the station. These ESDs must be easily accessible and should stop the dispenser and station operation and isolate hydrogen storage when actuated. The station must be designed such that it safely shuts down in the event of a fire, hydrogen gas detection, over-pressurization, ventilation system failure, activation of ESD, or abnormal fueling. Reactivation of the station from shutdown mode should only be through a manual restart by trained personnel.[1,2] This will prevent the station from restarting automatically

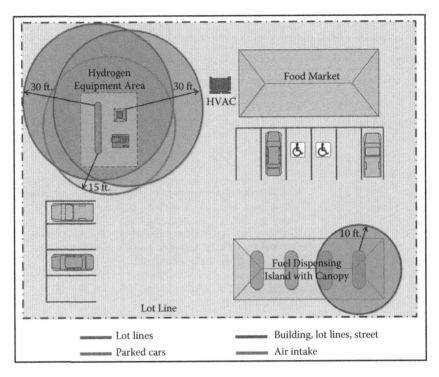

FIGURE 11.1
Example of setback distances for a hydrogen fueling station.

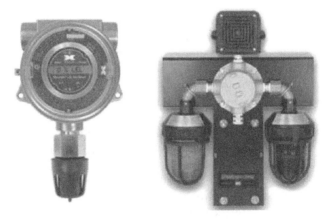

FIGURE 11.2
Gas detector and alarm station.

after the potential hazard has disappeared and will avoid the chances of the event happening again. For example, the hydrogen station may restart automatically after the concentration of hydrogen near the gas detector falls below its activating point and cause further hydrogen leakage.

Open flames or other potential source of ignition must not be allowed near the hydrogen equipment. The station and hydrogen equipment should be inspected and maintained periodically by trained personnel. Safety devices such as the pressure relief devices/valves, switches, transmitters, and detectors should be checked and calibrated periodically. Data logs, maintenance records, and other station information must be kept so that any safety issues or events can be readily identified and analyzed.

Even though the safety features mentioned above apply to both outdoor and indoor fueling stations, additional safety measures are required for indoor fast-filling facilities. All the compression, gas processing, and storage equipment should be located outside to reduce the amount of hydrogen release inside the facility. The area within 15 ft of the dispenser is classified as Class 1, Division 2, and Group B location and must be constructed with noncombustible or limited combustible material having a fire resistance of not less than 2 hours.[1,2] This area should also have a fire alarm system, fire sprinkler system, fire detection system, and gas detection system.

Ventilation is extremely important in indoor facilities to prevent hydrogen accumulation and avoid possible dangerous circumstances. According to NFPA codes, ventilation should be through a continuous mechanical ventilation system or by a mechanical ventilation system activated by a gas detection system when a hydrogen gas concentration of more than one quarter of the lower flammable limit (LFL) is detected inside the facility.[1,2] The gas detection system should also sound an alarm and provide visual indication when a maximum of 25% of the LFL is detected.[1,2] These measures will limit the chances of hydrogen being collected inside the room and the associated risk of fire or explosion. This would also reduce the chances of asphyxiation due to the displacement of oxygen inside the facility.

Ventilation rates are dependent on the size of the facility and should be at least 1 ft³/min · ft² (0.3 m³/min · m²) of the room area or more than 1 ft³/min · 12 ft³ (0.03 m³/min · 0.34 m³) of room volume.[1,2] However, according to NFPA 2[1] and NFPA 52,[2] a ventilation system is not required in indoor facilities where the following are satisfied:

- Minimum room volume based on maximum fueling event (Table 11.3)
- Ceiling height is greater than 25 ft
- Maximum flow rate less than or equal to 2 kg/minute
- Potential leak points must be monitored continuously
- Room size adjusted for number of dispensers
- Maximum length of the fueling nozzle is limited to 25 ft

TABLE 11.3

Minimum Room Volume Based on Maximum Fueling Event

Maximum Fuel Quantity per Dispensing Event		Minimum Room Volume	
lb	kg	ft³	m³
Up to 1.8	Up to 0.8	40,000	1,000
1.8 to 3.7	>0.8 to 1.7	70,000	2,000
3.7 to 5.5	>1.7 to 2.5	100,000	3,000
5.5 to 7.3	>2.5 to 3.3	140,000	4,000
7.3 to 9.3	>3.3 to 4.2	180,000	5,000

Source: Reprinted with permission from NFPA 2-2011, *Hydrogen Technologies Code.* Copyright © 2010, National Fire Protection Association, Quincy, MA. This reprinted material is not the complete and official position of the NFPA on the referenced subject, which is represented only by the standard in its entirety.

Note: Hydrogen quantity and room volume listed are from NFPA 2 2011 Edition.

For additional safety, the piping inside the facility should be welded construction with minimum connections to reduce the number of failure points. The piping pressure should be monitored continuously when the system is in idle mode to check for hydrogen leakage. Hydrogen venting from the dispenser should be done outside, away from any air intakes. Furthermore, there should be a manual shutoff valve and ESD located outside the facility to stop hydrogen flow and isolate hydrogen storage from the dispenser. This provides redundancy to the system safety and will stop hydrogen flow in the event that the shutoff valve and ESD located inside the facility are not accessible due to fire or hydrogen leakage. The safety measures discussed above will avert potentially dangerous conditions and will reduce the chance of hydrogen ignition in case of a leakage or release.

11.4 Hydrogen Vehicle Research and Development Garage

Since the infrastructure to support hydrogen vehicles is relatively new, the codes and standards for safe operation are still in the developing stage and the industry standards for research and development facilities for hydrogen vehicles are being formulated. The hydrogen research and development facility must be designed such that it can operate safely even in a hydrogen-rich environment. Hydrogen facilities require intrinsic safety barriers for electrical equipment to prevent spark and consequent ignition of hydrogen. Hence, all equipment in the hydrogen research and development facility must be rated for Class 1, Division 2, Group B hazardous location. The garage must also have sensors to detect the presence of hydrogen to determine if the concentration is approaching flammable concentration. For safely integrating and testing the hydrogen vehicle, the garage should have at least (1) a hydrogen gas detection system with alarms, (2) hazardous location electrical service, heating, cooling, ventilation, and lighting, and (3) emergency backup electric power system.

The construction, ventilation, and gas detection requirements of the garage are similar to the indoor fueling facilities. The garage should be constructed with noncombustible material or limited combustible and should have ventilation and gas detection system similar to that of the indoor fueling station. Since the hydrogen vehicle will be parked inside the garage, there is a chance of hydrogen leakage from the compressed hydrogen fuel tanks. Gaseous hydrogen, being lighter than air, will rise to the ceiling and can form an explosive mixture. Hence, the hydrogen detector should be located at the highest point inside the garage to readily detect the presence of hydrogen. It should also have a fire sprinkler system and the hydrogen piping should be in accordance to NFPA 55.[3]

Equipment such as parts washer, computers, car lift, and power tools is necessary for integrating and testing the vehicle. However, the equipment should be located in a separate area and used in strict compliance with the codes and standards. Measures should be taken to make it intrinsically safe—for example, Y-purged personal computer containment system. Open flames or other potential sources of ignition must not be allowed inside the garage. A handheld hydrogen detector will be very useful to detect small quantities of leaked hydrogen in the vehicle system. It is important that the persons working in the garage understand the properties of hydrogen and receive hydrogen safety training. Additional safety measures include warning signs like "WARNING—NO SMOKING—FLAMMABLE GAS" and "NON-ODORIZED GAS" near all access doors of the garage. All these measures will reduce the chances of fire, explosion, asphyxiation, or other potential hazardous.

11.5 Gaseous Hydrogen Storage

All hydrogen cylinders must be stored in compliance with NFPA 55.[3] The hydrogen gas cylinders must be safely secured, grounded, and protected from physical damage. Cylinders must be marked correctly with cylinder contents and maximum allowable working pressure of the container. This will prevent improper use and over-pressurization of the cylinder. Hydrogen storage areas should be cool and dry and have adequate ventilation to prevent the accumulation of hydrogen. The area within 15 ft of the hydrogen storage cylinders is classified as Class 1 Division 2 Group B location.[3] Appropriate pressure relief devices or valves must be installed to prevent over-pressurization and catastrophic failure of the cylinder. Hydrogen must be stored away from other combustible and readily ignitable material. Appropriate warning signs must be placed near the facility and potential sources of ignition such as smoking and open flames must not be allowed near storage areas. If hydrogen storage is connected to other systems, such as a hydrogen dispenser or a fuel cell, manual shutoff valves of other methods to isolate the system must be installed. The maximum amount of hydrogen that can be stored inside a building is 3500 standard cubic feet.[3] Asphyxiation is a major concern in indoor storage facilities, and such facilities should have hydrogen gas detectors with alarms, fire sprinkler system, and mechanical ventilation. Table 11.4 gives the minimum distances from outdoor gaseous hydrogen systems to exposures.

TABLE 11.4

Minimum Distances from Outdoor Gaseous Hydrogen Systems to Exposures

	Exposure	>250 to ≤ 3000 psi 0.747 in. ID (ft)	>3000 to ≤ 7550 psi 0.312 in. ID (ft)
1	Lot lines	45	30
2	Exposed persons other than those involved in servicing of the system	25	15
3	Buildings and structures		10
	Combustible construction	20	10
	Noncombustible non-fire-rated construction	20	5
	Fire-rated construction with a fire resistance rating of not less than 2 hours	5	
4	Openings in buildings of fire-rated or non-fire-rated construction (doors, windows, and penetrations)		
	Openable; fire-rated or non-fire-rated	45	30
	Unopenable; fire-rated or non-fire-rated	20	10
5	Air intakes (HVAC, compressors, other)	45	30
6	Fire barrier walls or structures used to shield the bulk system from exposures	5	5
7	Unclassified electrical equipment	15	15
8	Utilities (overhead), including electric power, building services, hazardous materials piping	20	10
9	Ignition sources such as open flames and welding	45	30
10	Parked cars	25	15
11	Flammable gas storage systems, including other hydrogen systems above ground		
	Non-bulk (<400 scf for hydrogen)	20	10
	Bulk	15	15

TABLE 11.4 (CONTINUED)

Minimum Distances from Outdoor Gaseous Hydrogen Systems to Exposures

	Exposure	>250 to ≤ 3000 psi 0.747 in. ID (ft)	>3000 to ≤ 7550 psi 0.312 in. ID (ft)
12	Aboveground vents or exposed piping and components of flammable gas storage systems, including other hydrogen systems below ground; gaseous or cryogenic	20	10
13	Hazardous materials (other than flammable gases) storage below ground; physical hazard materials or health hazard materials	20	10
14	Hazardous materials storage (other than flammable gases) above ground; physical hazard materials or health hazard materials	20	10
15	Ordinary combustibles, including fast-burning solids such as ordinary lumber, excelsior, paper, and combustible waste and vegetation other than that found in maintained landscaped areas	20	10
16	Heavy timber, coal, or other slow-burning combustible solids	20	10

Note: Reprinted with permission from NFPA 2-2011, *Hydrogen Technologies Code.* Copyright © 2011, National Fire Protection Association, Quincy, MA. This reprinted material is not the complete and official position of the NFPA on the referenced subject, which is represented only by the standard of its entirety.

11.6 Stationary Fuel Cell Installation

Stationary fuel cell applications are becoming increasingly attractive due to various incentive programs available to customers. They can be installed indoors or outdoors depending on the size and type of fuel cell and have high efficiency and reliability; low noise, maintenance, and environmental impact; and combined heat and power capability. The major safety concerns during the design, installation, and operation of a fuel cell are fuel supply and storage, electrical connections, and ventilation. Figure 11.3 shows the typical installation requirements for a stationary fuel cell in a commercial building.

1. Foundation and Protection
2. Fire Protection Systems
3. Piping Components and Connections

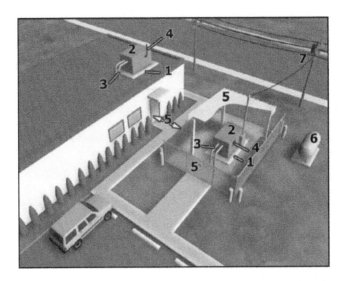

FIGURE 11.3
Typical installation requirements for a fuel cell in a commercial building. (From the U.S. Department of Energy, Energy Efficiency and Renewable Energy, Module 1, Permitting Stationary Fuel Cell Installations, Version 1.0, PNNL-14518, 2004.)

    4. Ventilation, Exhaust, and Makeup Air

    5. Siting, Installation, and Protection

    6. Fuel Supply and Storage

    7. Interconnections

If the fuel cell is installed indoors, it is recommended that a hydrogen detection system, connected to an emergency shutdown system, is also installed to detect any presence of hydrogen within the building. The hydrogen storage should be in accordance with NFPA 55[3] and should be outside whenever possible to avoid risk of hydrogen leaks inside the building. Additional emergency response actions should be incorporated into the existing building emergency response plan. All vents from the fuel cell system should be directed outside away from air intakes. All piping and wiring to and from the system should be clearly marked so that operators, emergency responders, or maintenance personnel can readily identify hydrogen and power lines. The fuel cell might be generating power even though the electric grid does not have power. Therefore, a separate electrical disconnect switch should be installed to isolate the fuel cell system from rest of the electrical circuit. Table 11.5 shows the codes and standards for stationary fuel cell installations.

TABLE 11.5

Codes and Standards Applicable to Stationary Fuel Cell Installations

#	Codes and Standards
1	International Mechanical Code (IMC)
2	International Fuel Gas Code (IFGC)
3	International Fire Code (IFC)
4	International Residential Code (IRC)
5	International Building Code (IBC)
6	International Plumbing Code (IPC)
7	NFPA 70—National Electric Code
8	NFPA 50A—Standard for Gaseous Hydrogen Systems at Consumer Sites
9	NFPA 50B—Standard for Liquefied Hydrogen Systems at Consumer Sites
10	NFPA 54—National Fuel Gas Code
11	NFPA 58—Liquefied Petroleum Gas Code
12	NFPA 853—Standard for the Installation of Stationary Fuel Cell Power Systems
13	AMSE Boiler and Pressure Vessel Code
14	UL 1741—Standards for Inverters Converters and Controllers for Use in Independent Power System
15	ANSI Z 21.83—Standard on Stationary Fuel Cell Power Plants

Source: Permitting Stationary Fuel Cell Installations, Module 1, U.S. Department of Energy, Energy Efficiency and Renewable Energy Version 1.0, PNNL-14518, 2004.

11.7 Conclusions

Hydrogen has many properties that make it unique including wide flammability limits, low ignition energy, high diffusivity, and low flame visibility. With proper understanding of these properties, incorporating experience, and safe handling procedures, hydrogen can be used in a safe working environment. Each facility has different risks associated with it and the potential hazards should be identified before deploying the system. A hydrogen-specific emergency response plan must be prepared and incorporated into the facility's emergency response plan. The operators must ensure that the local emergency responders are aware of this plan and should provide training if necessary.

This chapter contains hydrogen safety information for a hydrogen fueling station, hydrogen vehicle garage, hydrogen storage, and stationary fuel cell installation. However, readers should be aware that additional codes and standards may be applicable to these facilities. It is highly recommended that the readers verify the latest editions of the codes and standards being used. Furthermore, additional local codes and standards may also apply in each case.

References

1. NFPA 2—Hydrogen Technologies Code.
2. NFPA 52—Vehicular Gaseous Fuel Systems Code.
3. NFPA 55—Compressed Gases and Cryogenic Fluids Code.

12

Hydrogen Fuel Cell Vehicle Regulations, Codes, and Standards

Carl H. Rivkin

CONTENTS

12.1 Introduction

This chapter covers regulations, codes, and standards (RCS)[1] for hydrogen fuel cell vehicles. The chapter covers both domestic vehicle standards found primarily in Society of Automotive Engineers (SAE) and CSA Standards (CSA) documents and international standards found primarily in International Organization for Standardization (ISO) standards. The chapter does not cover the motor vehicle safety regulations promulgated by federal transportation safety agencies outside of the United States. The basic purpose of these RCS is to ensure safe operation of fuel cell–powered vehicles. These RCS do not cover the infrastructure required to support these vehicles. The infrastructure requirements are well developed in the United States, but they are outside of the scope of this chapter.

12.2 Hydrogen Fuel Cell Vehicle Deployment in the United States

There are several factors pushing the implementation of hydrogen fuel cell vehicles in the United States and in other, primarily industrialized countries. Hydrogen fuel cell vehicles do not produce hydrocarbon emissions and the hydrogen required to power their fuel cells could be produced using renewable energy technologies. An example of this process is the wind to hydrogen project at NREL, where wind turbines produce hydrogen by converting the electricity generated by the turbines to hydrogen through the use of an electrolyzer. The hydrogen is compressed to high pressures, in some cases over 5000 psi, and dispensed to hydrogen-powered vehicles.[2]

In the late 1990s the automobile industry Original Equipment Manufacturers (OEMs) became interested in the potential development of hydrogen fuel cell vehicles. Their interest led to the DOE projects designed to further the development and deployment of hydrogen fuel cell vehicles. This deployment effort included the development of the RCS required to deploy hydrogen fuel cell vehicles on a commercial scale. This commercial deployment has been projected to take place in the 2015–2020 time frame, although at one time the U.S. Department of Energy (DOE) had projected an earlier commercial deployment. Figure 12.1 shows a DOE estimated timeline for a transition to widespread use

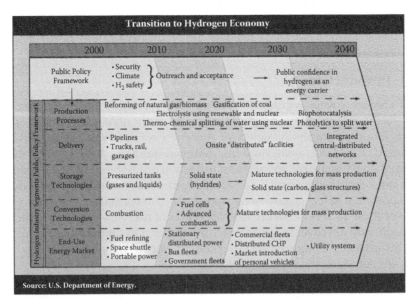

FIGURE 12.1
DOE timeline for the deployment of hydrogen fuel cell vehicles. (From the U.S. Department of Energy, Office of Energy Efficiency and Renewable Energy, http://www1.eere.energy.gov/hydrogenandfuelcells/mypp/.)

of hydrogen that would include deployment of hydrogen fuel vehicles. The RCS effort would be included in the row titled "Public Policy Framework." The intent when this timeline was developed was that RCS would be addressed by 2010.

Current OEM projections vary by company, but several OEMs plan on introducing hydrogen fuel cell vehicles in the 2015–2020 time frame. The expectation is that RCS would be in place to allow for commercial deployment of road vehicles by 2015.

Industry, DOE, and other interested parties have participated in this effort to develop comprehensive RCS to allow for commercial deployment of hydrogen fuel cell vehicles.

As work progressed in developing road fuel cell vehicles and fuel cells were developed that could effectively power vehicles, another promising hydrogen fuel cell application appeared: hydrogen fuel cell–powered forklift vehicles or, as they are more generally categorized, industrial trucks. Although the RCS for many of the infrastructure requirements for hydrogen fuel cell–powered forklifts are the same as hydrogen fuel cell road vehicles, there are new standards required to address the hydrogen fuel cell forklift vehicle and standards specific to fueling a forklift vehicle.

12.3 Overview of RCS

Table 12.1 lists RCS that apply to hydrogen fuel cell vehicles by organization. The table is not all-inclusive but lists key RCS used both domestically and internationally.

Figure 12.2 shows progress made on several RCS issues including outreach and NREL efforts to directly assist in deployment through third-party safety reviews.

12.4 Road or Passenger Hydrogen Fuel Cell Vehicles

Figure 12.3 shows the basic components of hydrogen fuel cell–powered road vehicle.

The RCS, particularly standards, have been focused on the fuel cell stack and the hydrogen storage system because these are the systems in the vehicle that store and process hydrogen. Standards have been developed or are being developed that address both the system level and the component level safety issues associated with the hydrogen storage and fuel cell stack.

TABLE 12.1

Regulations, Codes, and Standards for Hydrogen Fuel Cell Vehicles

Regulation, Code, or Standard	Subject Matter	Enforcing Authority
1. CSA onboard road vehicle component standards http://www.csa-america.org/standards/current_activites/default.asp?load=alternative	• CSA America HGV 2 (draft) • CSA America HGV 3.1 (draft) Fuel System Components for Hydrogen Gas Powered Vehicles • CSA America HGV 4.1 (TIR) Hydrogen Gas Dispensing Systems • CSA America HGV 4.2 (TIR) Hoses for Hydrogen Gas Vehicles and Dispensing Systems • CSA America HGV 4.3 (draft) Temperature Compensation Devices for Hydrogen Gas Dispensing Systems • CSA America HGV 4.4 (TIR) Breakaway Devices for Hydrogen Gas Dispensing Hoses and Systems • CSA America HGV 4.5 (TIR) Priority and Sequencing Equipment for Hydrogen Gas Dispensing Systems • CSA America HGV 4.6 (TIR) Manually Operated Valves for Hydrogen Gas Dispensing Systems • CSA America HGV 4.7 (TIR) Automatic Valves for Use in Hydrogen Gas Vehicle Fueling Stations • CSA America HGV 4.8 (TIR) Hydrogen Gas Fueling Station Reciprocating Compressor Guidelines • CSA America HPRD 1 (draft) Pressure Relief Devices for Hydrogen Gas Vehicle (HGV) Containers	OEM due diligence enforcement
2. CSA onboard industrial truck standards http://www.csa-america.org/standards/current_activites/default.asp?load=alternative	CSA HPIT 1 Compressed Hydrogen Powered Industrial Truck Onboard Fuel Storage and Handling Components and (under development) CSA HPIT 2 Compressed Hydrogen Station and Components for Fueling Powered Industrial Trucks (under development)	U.S. Occupational Safety and Health Administration (OSHA)

TABLE 12.1 (CONTINUED)

Regulations, Codes, and Standards for Hydrogen Fuel Cell Vehicles

Regulation, Code, or Standard	Subject Matter	Enforcing Authority
3. SAE onboard road vehicle system standards http://standards.sae.org/fuels-energy-sources/on-board-energy-sources/fuel-cells/standards/	SAE J2579, SAE J2600 • J1766 Recommended Practice for Electric and Hybrid Electric Vehicle Battery Systems Crash Integrity Testing • J2572 Recommended Practice for Measuring Fuel Consumption and Range of Fuel Cell and Hybrid Fuel Cell Vehicles Fuelled by Compressed Gaseous Hydrogen • J2574 Fuel Cell Vehicle Technology • J2578 Recommended Practice for General Fuel Cell Vehicle Safety • J2579 Fuel Cell Systems in Fuel Cell and Other Hydrogen Technologies • J2594 Recommended Practice to Design for Recycling Proton Exchange Membrane (PEM) Fuel Cell Systems • J2600 Compressed Hydrogen Surface Vehicle Refueling Connection Devices • J2615 Testing Performance of Fuel Cell Systems for Automotive Applications • J2616 Testing Performance of the Fuel Processor Subsystem of an Automotive Fuel Cell System • J2719 Information Report on the Development of a Hydrogen Quality Guideline for Fuel Cell Vehicles • J2760 Pressure Terminology Used in Fuel Cells and Other Hydrogen Vehicles	OEM due diligence enforcement
4. SAE road vehicle fueling standards http://standards.sae.org/fuels-energy-sources/on-board-energy-sources/fuel-cells/standards/	SAE J2601 Fueling Protocols for Light Duty Gaseous Hydrogen Surface Vehicles	OEM due diligence enforcement/Building and Fire officials

(*continued*)

TABLE 12.1 (CONTINUED)

Regulations, Codes, and Standards for Hydrogen Fuel Cell Vehicles

Regulation, Code, or Standard	Subject Matter	Enforcing Authority
5. SAE industrial truck fueling standards http://standards.sae.org/fuels-energy-sources/on-board-energy-sources/fuel-cells/standards/	SAE J2919 (under development)	OSHA/Building and Fire officials
6. Industrial truck performance standards http://ulstandardsinfonet.ul.com/tocs/tocs.asp?fn=2267.toc http://www.nfpa.org/catalog/product.asp?title=Code-505-2011-Fire-Safety-Powered-Industrial-Trucks&category%5Fname=Codes+and+Standards&pid=50511&target%5Fpid=50511&src%5Fpid=&link%5Ftype=category&order%5Fsrc=	1. **UL2267 Fuel Cell Power Systems for Installation in Industrial Electric Trucks** 2. **NFPA 505 Fire Safety Standard for Powered Industrial Trucks Including Type Designations, Areas of Use, Conversions, Maintenance, and Operations**	OSHA/Building and Fire officials
7. Federal Motor Vehicle Safety Standards (for fuel cell vehicles)	Not promulgated	U.S. Department of Transportation (DOT)
8. Global Technical Regulations (GTR)	Not promulgated	DOT
9. ISO standards http://www.iso.org/iso/standards_development/technical_committees/list_of_iso_technical_committees/iso_technical_committee.htm?commid=54560	ISO 13984:1999 Liquid hydrogen—Land vehicle fuelling system interface ISO 13985:2006 Liquid hydrogen—Land vehicle fuel tanks	Varies by country

TABLE 12.1 (CONTINUED)

Regulations, Codes, and Standards for Hydrogen Fuel Cell Vehicles

Regulation, Code, or Standard	Subject Matter	Enforcing Authority
	ISO/DIS 14687-2 Hydrogen fuel—Product specification—Part 2: Proton exchange membrane (PEM) fuel cell applications for road vehicles	
	ISO 16111:2008 Transportable gas storage devices— Hydrogen absorbed in reversible metal Hydride	
	ISO 17268:2006 Compressed hydrogen surface vehicle refueling connection devices ISO15869 Gaseous hydrogen and hydrogen blends—Land vehicle fuel tanks	

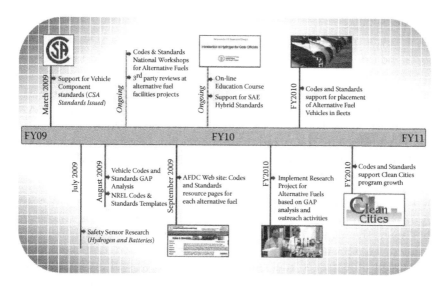

FIGURE 12.2
Snapshot of NREL efforts to support RCS development and vehicle deployment.

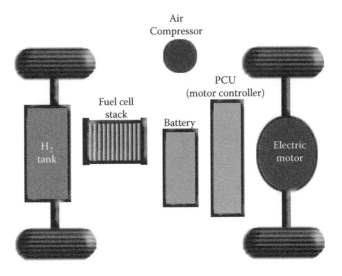

FIGURE 12.3
Basic components of a hydrogen fuel cell–powered vehicle. (From the U.S. Department of Energy, Office of Energy Efficiency and Renewable Energy, http://www.afdc.energy.gov/afdc/vehicles/fuel_cell_what_is.html.)

RCS for fuel cell road vehicles are covered at three levels:

1. Level 1: The international treaty level through Global Technical Requirements (GTR) that are under development for hydrogen fuel cell vehicles

2. Level 2: The national level through Federal Motor Vehicle Safety Standards (FMVSS) (that would be potentially developed in the next few years)

3. Level 3: The standards level through standards (both national and international) developed by such organizations as SAE, CSA, Underwriters Laboratories (UL), International Organization for Standardization (ISO), and the National Fire Protection Association (NFPA). OEMs comply with these standards through what in many cases is a voluntary compliance system.

The third level (standards) is where most work has been done. Given that commercial deployment of hydrogen fuel cell road vehicles is still several years away, the standards level of RCS is where most work has been concentrated. FMVSS would typically not be promulgated until commercial deployment of hydrogen fuel cell vehicles. The GTR will potentially make international deployment of hydrogen fuel cell vehicles occur more smoothly but is not necessary for deployment of vehicles in the United States.

Although it is not the subject of this chapter, there has been an extensive effort to develop RCS for infrastructure required to support hydrogen fuel cell vehicles. This chapter covers the vehicle RCS. These requirements can be generally divided into system-level requirements and component requirements.

12.5 Standards for Fuel Cell Road Vehicles

The SAE is developing requirements for hydrogen fuel cell systems on vehicles. Two of the key documents that they are developing are SAE J2579 (Fuel Cell Systems in Fuel Cell and Other Hydrogen Technologies) and SAE J2601 (Fueling Protocols for Light Duty Gaseous Hydrogen Surface Vehicles).

The purpose of SAE J2579 is to define design, construction, operational, and maintenance requirements for hydrogen storage and handling systems in road vehicles. SAE J2579 defines performance-based requirements for verification of design prototype and production hydrogen storage and handling systems. Test protocols (for use in type approval or self-certification) to qualify designs (and/or production of systems) as meeting the specified performance requirements are described. Crashworthiness of hydrogen storage and handling systems is beyond the scope of SAE J2579. SAE J2578 includes requirements relating to crashworthiness and vehicle integration for fuel cell vehicles. SAE J2578 establishes industry performance criteria for vehicle crashes. Legally compliant design qualification for crash impact resistance is achieved by demonstrated compliance of the vehicle with applicable FMVSS regulations.

Hydrogen fuel cell road vehicles generally will use lighter than stainless steel composite tanks. Vehicle tanks are categorized in ISO 15689 as Type 1–4. Type 4 is composite wrapped tanks. Table 12.2 shows standards for high-pressure hydrogen tanks.

SAE Temporary Interim Requirement (TIR) J2601 establishes safety limits and performance requirements for gaseous hydrogen fuel dispensers. The criteria include maximum fuel temperature at the dispenser nozzle, the maximum fuel flow rate, the maximum rate of pressure increase, and other performance criteria based on the cooling capability of the station's dispenser. This document establishes fueling guidelines for non-communication fueling. This is fueling where there is not communication that is transmitted from the vehicle and utilized at the dispenser to control the vehicle-fueling rate. The document also established guidelines for communication fueling where the information transmitted from the vehicle is used to modify the vehicle-fueling rate. The process by which fueling is optimized using vehicle-transmitted information is specified. This document provides details of the communication data transmission protocol. The mechanical connector geometry is not covered in this document. SAE J2600 defines the connector requirements for fueling vehicles operating with a nominal working pressure of 5000 psi. SAE TIR J2799 defines the mechanical connector geometry for fueling vehicles to 10,000 psi and also provides

TABLE 12.2

Current Standards Compliance for 25-(Megapascal) MPa, 35-MPa, and 70-MPa Pressure Vessels

Storage Pressure	Standards Compliance
25 MPa (3.6 ksi)	NGV2-2000 (modified)
	DOT FMVSS 304 (modified)
35 MPa (5 ksi)	E.I.H.P./Rev 12B
	ISO 15869 is derived from EU 97/23/EG
	NGV2-2000 (modified)
	FMVSS 304 (modified)
	Reijikijyun Betten 9
70 MPa (10 ksi)	E.I.H.P./Rev 12B
	ISO 15869 is derived from EU 97/23/EG
	FMVSS 304 (modified)
	Betten 9 (modified)

Source: U.S. Department of Energy, Office of Energy Efficiency and Renewable Energy, http://www1.eere.energy.gov/hydrogenandfuelcells/storage/hydrogen_storage_testing.html.

specifications for the hardware for vehicle-to-station dispenser communication. It is expected that SAE J2600 will be revised to include the receptacle content of SAE TIR J2799, in which case the resulting SAE J2600 will provide connector hardware requirements for gaseous hydrogen fueling at all working pressures.

SAE J2601 will likely be revised to include separate requirements for fueling heavy-duty vehicles and motorcycles as well as forklifts and for residential hydrogen fueling appliances. There is significant difference between the onboard storage capacity of heavy-duty and light-duty vehicles. SAE J2601 applies to fueling using an average pressure ramp rate methodology that may need to be verified with a hydrogen dispenser test apparatus.

CSA HGV 4.3, Temperature Compensation Devices for Hydrogen Gas Dispensing Systems, *defines such a methodology. CSA HGV 4.3 includes provisions for optional alternative communications fueling protocols. New dispenser protocol proposals would need to be verified with data and experience demonstrating the fueling algorithm's capability to operate within the constraints of Section 5 of SAE J2601.*

The following are three of the key documents that CSA is working on:

1. HPRD 1 2009 edition, Temporary Interim Requirements (TIR) for pressure relief devices for vehicle containers
2. HGV 4.2, Hoses and Hose Assemblies 2009 published as TIR
3. HGV 4.3, Fueling Station Fueling Parameter Verification (draft expected in early 2012)

CSA has been working with SAE to develop component and test standards that complements the SAE effort to develop SAEJ2601. The purpose of this HGV 4.3 document is to do the following:

(1) Have vehicles meet the requirements (i.e., generally prescriptive) in **J2601** for safe fueling without:

　　(i)　Over-pressuring;

　　(ii)　Exceeding temperature specifications; and

　　(iii)　Exceeding density requirements

(2) Meet performance-based requirements for the safe fueling of vehicles without:

　　(i)　Over-pressuring;

　　(ii)　Exceeding temperature specifications; and

　　(iii)　Exceeding density requirements—whether the requirements are provided in a document other than J2601 or developed by OEMs

12.6　Standards for Hydrogen Fuel Cell Industrial Trucks

Figure 12.4 shows a hydrogen fuel cell–powered industrial truck. This unit was originally battery-powered but was converted to run on a fuel cell stack. These conversions can be done without extreme retrofitting of the forklift.

FIGURE 12.4
Hydrogen fuel cell–powered forklift. (Photograph taken by Carl Rivkin, 2008).

The concerns about weight reduction in road vehicles do not apply in fork-lifts because forklifts need weight to counterbalance the load that they lift. Type 1 metal tanks can be used in forklifts. These tanks would typically not be used in a passenger vehicle because of their weight.

There are several standards that apply specifically to industrial hydrogen fuel cell industrial trucks or industrial trucks generally. These documents include the following:

1. UL2267, Fuel Cell Power Systems for Installation in Industrial Electric Trucks
2. NFPA 505, Fire Safety Standard for Powered Industrial Trucks Including Type Designations, Areas of Use, Conversions, Maintenance, and Operations
3. CSA HPIT (Draft), Compressed Hydrogen Powered Industrial Truck: On-board Fuel Storage and Handling Components
4. CSA HPIT 2 (Draft), Compressed Hydrogen Station and Components for Fueling Powered Industrial Trucks

12.7 Ongoing RCS Development Issues

The GTR is expected to produce a Phase I document in the 2011/2012 time frame.

However, this issuance is not impeding deployment of road fuel cell vehicles. Major RCS issues that more directly affect the deployment of road fuel cell vehicles are as follows:

1. Completing work on SAE J2579 to define acceptable storage system performance criteria
2. Issuing SAE J2601 that defines high-pressure fueling protocols as a final document
3. Completing SAE J2719 that defines acceptable contaminant levels in hydrogen as a fuel for proton exchange membrane (PEM) fuel cell vehicles
4. Completing CSA HGV 4.3 that validates fueling systems

There are also ongoing projects that address hydrogen fuel cell–powered industrial trucks. These include the following:

1. Ongoing work by CSA to develop standards for fuel systems and components for hydrogen fuel cell–powered industrial trucks
2. Ongoing work to update NFPA 505 to reflect new hydrogen fuel cell applications

Because most fuel cell vehicles will be sold in more than one country, coordination of domestic and international standards is a concern. Much of this coordination is build into the standard development process. For example, SAE standards are used in several countries and the technical committees that develop the SAE standards have representation from multiple countries and both domestic and international companies.

Some standards are being developed in parallel. For example ISO 14687-2 (Hydrogen fuel—Product specification—Part 2: Proton exchange membrane [PEM] fuel cell applications for road vehicles) and SAE J2719 are being developed in parallel so that domestic and international fuel quality standards are not in conflict.

12.8 Conclusions

RCS are in place that allow the deployment of prototype road hydrogen fuel cell vehicles and commercial hydrogen fuel cell–powered industrial trucks.

The RCS for road vehicles require the promulgation of FMVSS for widespread commercial deployment. There are also several vehicle standards under development that need to be issued as final documents for this widespread deployment to occur. The deployment of hydrogen-powered industrial trucks does not require FMVSS, and although there are several standards under development that cover systems and components on these vehicles, they are not impeding the deployment of these vehicles.

References

1. RCS, http://www.afdc.energy.gov/afdc/codes_standards.html.
2. National Renewable Energy Laboratory, http://www.nrel.gov/hydrogen/proj_wind_hydrogen.html.

Index

Printed and bound by CPI Group (UK) Ltd, Croydon, CR0 4YY

18/10/2024

01776256-0011